Optical Sensors for Biomedical Diagnostics and Environmental Monitoring

Optical Sensors for Biomedical Diagnostics and Environmental Monitoring

Banshi Dhar Gupta
Anand Mohan Shrivastav
Sruthi Prasood Usha

CRC Press
Taylor & Francis Group
Boca Raton London New York

CRC Press is an imprint of the
Taylor & Francis Group, an **informa** business

CRC Press
Taylor & Francis Group
6000 Broken Sound Parkway NW, Suite 300
Boca Raton, FL 33487-2742

International Standard Book Number-13: 978-1-4987-8906-6 (Hardback)

Library of Congress Cataloging-in-Publication Data

Names: Gupta, Banshi Dhar, author. | Shrivastav, Anand Mohan, author. | Usha, Sruthi Prasood, author.
Title: Optical sensors for biomedical diagnostics and environmental monitoring / Banshi Dhar Gupta, Anand Mohan Shrivastav, Sruthi Prasood Usha.
Description: Boca Raton, FL : CRC Press, Taylor & Francis Group, [2017] | Includes bibliographical references and index.
Identifiers: LCCN 2017023196| ISBN 9781498789066 (hardcover ; alk. paper) | ISBN 1498789064 (hardcover ; alk. paper) | ISBN 9781315156033 (eBook) | ISBN 1315156032 (eBook) | ISBN 9781498789073 (eBook) | ISBN 1498789072 (eBook) | ISBN 9781351652100 (eBook) | ISBN 1351652109 (eBook) | ISBN 9781351642569 (eBook) | ISBN 1351642561 (eBook)
Subjects: LCSH: Optical fiber detectors. | Environmental monitoring. | Plasmons (Physics)
Classification: LCC TA1815 .G87 2017 | DDC 681/.25--dc23
LC record available at https://lccn.loc.gov/2017023196

Visit the Taylor & Francis Web site at
http://www.taylorandfrancis.com

and the CRC Press Web site at
http://www.crcpress.com

Contents

Preface

This book is an effort to bring a comprehensive book on optical sensors for biomedical diagnostics and environmental monitoring for the undergraduate and graduate students apart from researchers working in various related fields. The idea of writing this book came during a group discussion where it was felt that a detailed information on various kinds of sensors realized on this topic incorporating the newly implemented and upcoming techniques should be provided to the community because of their direct relation to the health of the living body like human beings, animals, etc. Various kinds of newly developed techniques are being used for the development of sensors for such applications. Surface plasmon resonance (SPR) is one of such techniques which has been widely used and can be easily understood with the elementary knowledge of electromagnetic theory and hence can be recommended to the students with its possible applications. The literature has also shown the upcoming technique of lossy mode resonance (LMR) for sensing applications. LMR exhibited by the semiconductor metal oxide (SMO) in the visible region and possessing applications in sensing of various gases, chemical and biological analytes also need wide attention. The invention of optical fiber technology has made easy implementation of SPR- and LMR-based sensors with the unique advantages of flexible designs with capability of online monitoring and remote sensing using cost-effective substrate. Nanotechnology is an intense branch that has made its role clear in sensing applications by its irreplaceable performances in many aspects. However, the integration of optical fiber and nanotechnology has fascinated the world of the sensors. Therefore, a comprehensive collective knowledge of this integration and its advantages in the development of sensors is required. The very new field of molecular imprinting is gaining its attraction mainly in the field of biomolecular recognition. This book is the first work to explain in detail the usage of this method mainly in fiber-optic sensors and, in particular, in integrating it with the nanoparticles for recognizing biosensors. With this book, we aim to introduce the graduate students and early researchers to an overview of the plasmonic and LMR-based sensors for biomedical diagnostics and environmental monitoring employing the miracles of the fascinating fields, their working principles, and applications.

This book will benefit the researchers as well as the industrial and public society with an overview of recent advances of optical sensors and their applications for human health and safety. This book starts with the introduction of sensors, need of sensors, and distinct types of sensors applicatory for biomedical and environmental monitoring (Chapter 1). Biomedical applications of sensors are gaining importance intently and hence a deep attention is given to biosensors for the diagnosis of various diseases important for healthcare in this chapter. Chapter 2 introduces the basic concepts of resonance and various resonance-based sensing techniques that are currently being used in various research and industrial applications. SPR and its collateral techniques are discussed in detail along with the chronological development of the field. In addition, a new technique, LMR, developing its application in the sensing, is also explored. In Chapter 3, a discussion about the fiber-optic sensors, the essential components of fiber-optic sensors, their functions, and the parameters tailoring sensor performance are presented. The multiple design possibilities such as D-shaped, tapered and U-shaped miniaturized probe, in addition to the uniform straight core probe, are included. This is followed by the discussion on nanostructures and their role as a sensing platform in Chapter 4. The chapter includes various metallic nanostructures useful in sensing and their fabrication methods. A number of sensors utilizing nanostructures are presented for environmental and agricultural monitoring, biological applications, biomarkers, surgical, and clinical diagnostics. In Chapter 5, an analysis on the action of SMO for sensing purpose is presented. The section illustrates the SPR and LMR mechanism of sensing, involving metal oxide, for the monitoring of gases, their upcoming applications in biomolecular detection and diagnosis. Chapter 6 introduces the latest technique of molecular imprinting incorporating the qualitative analysis and amelioration done by the technique in the field of sensing applications. The sensors realized using molecular imprinting for food safety, environmental monitoring, and biomedical usage are discussed in the chapter. The last chapter of this book (Chapter 7) summarizes all the contents with a note on future trends of research in the field of fiber-optic and other technique-based sensors for environmental monitoring and biomolecular recognition. It is hoped that this book will serve the purpose for which it has been written.

Banshi Dhar Gupta
Anand Mohan Shrivastav
Sruthi Prasood Usha
Physics Department
Indian Institute of Technology Delhi
New Delhi, India

Acknowledgments

I consider myself very fortunate to have worked with Dr. Chandra Deep Singh, Professor Sunil Khijwania, and Dr. Navneet K. Sharma on fiber-optic evanescent field absorption sensors. My continuous interaction with them has made me wiser in the field of fiber-optic sensors. It was this interaction that resulted in the publication of my first book titled *Fiber Optic Sensors: Principles and Applications* (2006). In 2004, I took an initiative to work on fiber-optic sensors utilizing surface plasmon resonance technique. The research carried out on this topic resulted in 10 PhD theses. I thank Drs. Anuj K. Sharma, Rajan Jha, Rajneesh K. Verma, Sachin K. Srivastava, Sarika Singh, Priya Bhatia, Roli Verma, Satyendra K. Mishra, and Rana Tabassum who have worked with me in this exciting field. A part of their work resulted in my second book titled *Fiber Optic Sensors Based on Plasmonics* (2015). For the last 5 years, my focus has been on the research and development of optical sensors for biomedical diagnostics and environmental monitoring. My present graduate students, apart from Anand Mohan Shrivastav and Sruthi Prasood Usha (also the authors of this book), Ravi Kant, Vivek Semwal, Anisha Pathak, and Sonika Sharma are working in this area. I learnt a lot from these colleagues through interaction on this topic. Indeed, a part of the book has emerged from the work carried out with these graduate students and with some of my old students. I thank all of them for their hard work in the laboratory without which publication of this book was not possible. Finally, special thanks to my wife Uma Gupta for her patience and understanding during the writing of this book.

Banshi Dhar Gupta

About the Authors

Banshi Dhar Gupta earned his MSc in physics in 1975 from Aligarh Muslim University, India and a PhD in physics in 1979 from the Indian Institute of Technology, New Delhi. In 1978 he joined the Indian Institute of Technology, New Delhi, where he is currently a professor of physics. In addition, Professor Gupta has worked at the University of Guelph, Canada between 1982 and 1983, the University of Toronto, Canada in 1985, the Florida State University, Florida, USA in 1988, the University of Strathclyde, UK in 1993, and the University of Birmingham, UK in 2010. In 1992, he was awarded the ICTP Associateship by the International Centre for Theoretical Physics, Trieste, Italy, which he held for 8 consecutive years. In this capacity, he visited ICTP, Italy in 1994 and 1996. Professor Gupta is a recipient of the 1991 Gowri Memorial Award of the Institution of Electronics and Telecommunication Engineers, India. He has published more than 170 research papers in international journals and 100 papers in conferences. Professor Gupta has authored books titled *Fiber Optic Sensors: Principles and Applications* (New India Publishing Agency—NIPA, New Delhi, 2006) and *Fiber Optic Sensors based on Plasmonics* (World Scientific, 2015), and is the coeditor of the *Proceedings* of SPIE, USA, Vol. 3666, 1998 and Vol. 8173, 2010. Professor Gupta has delivered invited talks on plasmonics-based sensors in various international conferences held in the United States, China, Australia, Korea, and India. Professor Gupta is a member of the editorial board of the *Journal of Sensors*. His current areas of interest are plasmonic biosensors, fiber optic sensors, and nanotechnology. He is a regular member of the Optical Society of America and life member of the Optical Society of India.

Anand Mohan Shrivastav earned his MSc in physics in 2012 from C.S.J.M. University Kanpur, India. He joined IIT Delhi as a full time PhD student in 2013. Mr Shrivastav has published more than 15 research papers in international journals and 10 papers in conferences. His current areas of interest are fiber optic sensors, nanotechnology, and molecular imprinting technology. Mr Shrivastav was awarded senior research fellowship in 2015 by the University Grants Commission (UGC), India. He is a member of the Optical Society of America and Optical Society of India.

Sruthi Prasood Usha earned her BTech in electronics and communication engineering in 2010 from PAACET (P.A. Aziz College of Engineering and Technology), Kerala University, India and MTech in optoelectronics and laser technology in 2013 from Cochin University of Science and Technology (CUSAT, India). Ms. Usha joined IIT Delhi as a full time PhD student in 2014. She has published 12 research papers in international journals and 10 papers in conferences. Her current areas of research are fiber optic SPR (surface plasmon resonance) and LMR (lossy mode resonance) sensors, nanostructures, and nanotechnology. She is a member of the Optical Society (OSA).

1

Introduction to Sensors

Measurements of various physical, chemical, and biological parameters/analytes such as pressure, temperature, magnetic field, pH, pesticides, gases, glucose, urea, cancer, etc. are required for various applications such as environment and health monitoring, process control, and electrical utilities/power plants. The devices used to measure these parameters are called sensors. Tremendous amount of work has been carried out on the development of low-cost, compact, sensitive, and analyte-specific sensors. Their progression over the decades has reached well-established pace due to the rapidly developing technology and science. The sensors are named based on the technique used by them for the measurement of abovementioned parameters. The technique can be optical, electrical, mechanical, electrochemical, or any other. The present book deals with the biomedical diagnostics and environmental monitoring sensors utilizing optical techniques. Below, we shall first define a sensor and then discuss the important features of a sensor such as need of a sensor, its components, performance parameters, and classifications.

1.1 What Is a Sensor?

Sensor is a device which detects and analyzes a parameter continuously by providing an output in a way that is readable by humans or human-controllable instruments in a continuous manner with possibilities of reproducibility and recovery. Further, in a sensor, change in one form of input energy to corresponding single-valued change in another or same form of energy takes place. The output of a sensor is usually electrical or optical in nature. The sensors of the current decade can quantify and qualify almost all the physical, mechanical, chemical, and biological entities. There are many things which are expected from a sensor. First, the sensor should be insensitive to the change in external environmental parameters such as temperature and humidity (for nontemperature and humidity sensors) to provide the invariable results. Second, the allocation of a sensor to analyze a particular parameter in a natural environment should be made considering its ability to withstand harsh atmosphere conditions and changes, the domain in which the sensor holds application, its lifetime, remote activities, and online compatibility with good recovery. Considering all these aspects, the work reported in the literature so far has concluded that the optical way of commutating changes is better than the electrochemical and calorimetric means. In optical sensors, the method of converting one form of energy to another is optical; however, the system may make use of other fields such as electrical, electronic, mechanical, biological, nanostructural, material, biochemical, chemical, and physical engineering along with optics. Thus, the versatile-integrative behavior of optical sensors covers a broad area of requirements.

The optical-sensing system, as in the case of all other methods, has four basic units. The foremost unit is the light source of the sensor that needs to be selected according to the design of the sensor system and the application to which it is oriented. The second unit is the sensor receptor/recognition element which interacts with the analyte (to be sensed) and changes its own optical properties such as refractive index. The change in the properties of recognition unit is determined by the third unit called as transducer. The transducer is the unit which converts the change in the recognition unit properties into some measurable signal by means of an optical transducing phenomenon. The measurable parameter of the signal can be the intensity, phase, or wavelength of the light source which is detected with the help of detector/analyzer which is the fourth and last unit of a sensor. Figure 1.1 shows the schematic of a generic sensor with all the required units. The working of the whole system depends on the mechanism as well as the principle on which the sensor works and also on its utilization. For a sensing system based on optical method, the variation in one of the properties of light such as amplitude/intensity, phase, polarization, frequency, and wavelength due to the change in the sensing medium is analyzed.

Optical-sensing system may use lens, prism, mirror, grating, microscope objective, waveguide, fiber, etc. and optical detection as well as analyzing units. The detection in optical system can be performed by schemes of luminescence, fluorescence, and absorption. The advantages of optical sensing include immunity to electromagnetic interference, possibilities of point and distributed arrangement with multiplexing, compatibility, and

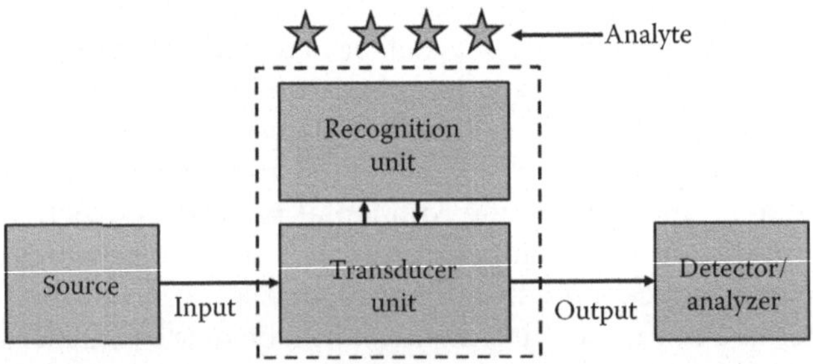

FIGURE 1.1
Schematic of a generic sensor with the essential units.

electrical isolation (Grattan and Sun 2000). The usage of fiber as the optical substrate competes with all other schemes of optical sensing, as well as sensing based on transducer elements other than optical due to its capability of remote sensing and online monitoring. Hence, in this book, we shall impart the significance of optical sensing, mainly using optical fiber, for the physical, chemical, and biological sensing to cover environmental supervision and biomedical diagnostic applications, which cover the relevant areas of sensor operation in present and future. The upcoming sections discuss the relevance of sensors in detecting various biological, gaseous–chemical measurands with the parameters which quantify the sensor performance. Section 1.5 will discuss the types of classifications of sensors followed by sensor regeneration aspects in Section 1.6. Finally, Section 1.7 contests the framework of the book.

1.2 Need of Sensors

The sensors are one of the most important devices that we use in day-to-day life. As we all know, our nervous system consists of five vital sensing organs: nose, skin, eyes, ears, and tongue. These are used for the sensing of olfacception (smell), tactioception (touch), ophthalmoception (sight), audioception (listening), and gustaoception (taste), respectively. Various types of devices and machines are equipped with sensors that help them in working. Consider a mobile phone, it is a device combined with multisensor technology such as accelerometer and gyroscope that are used for the determination of linear and angular rotations. Various types of temperature and humidity sensors are being used daily in several electronic machines such as refrigerators, air conditioner, and air cooler for the automated working. Light-dependent resistors (LDRs) are used as sensors in the automatic doors and streetlights with an ability to switch their operation. Apart from these, sensors are also used in various fields for the human and environmental safety, biomedical applications, military applications, etc. The aim of sensor development for the environmental monitoring is the safeguard of the living organisms from a number of pathogens and toxic contaminants that are continuously released into the environment, resulting in the pollution of air, soil, and water (Rodriguez-Mozaz et al. 2004). Sensors are also required in various biomedical and clinical applications to detect glucose, urea, vitamins, and nutrients concentrations in human body, and also need to be detected for ensuring their level within the normal range in the body. In order to ensure the safety of food and water, several types of sensors for the detection of impurities such as heavy metal ions (to secure for the use of hard drinking water), pesticides, melamine, etc. are used and further advancements in these sensors are required for highly specific and selective detection. In accordance with the applications of sensors for biomedical, clinical, environment, and food safety, the sensors can be classified in three broad categories as biosensors, gas sensors, and chemical sensors.

1.2.1 Biosensors

Biosensor is a device which is generally designed and fabricated with a biological sensing element incorporating a transducer. It generates a change in the transducer signal or bio-recognition element characteristics on exposure to an analyte for which it has been designed. The change in signal depends upon the concentration of the analyte. A number of definitions have been given to the biosensor by the people working in the field of biosensing. The general definition is: "Biosensor is a device that utilizes biological components such as enzymes, antigens/antibodies for the determination of the concentration of the biomaterials" (Wang and Wolfbeis 2013). The biomaterials to be detected are usually the materials which are toxic in nature and can harm human body as well as the environment. The invention of first biosensor was by Leland Clark and hence he was renowned as the father of the biosensing concept (Clark and Lyons 1962). He showed detection of glucose using oxygen electrode, where glucose oxidase was entrapped over a Clark oxygen electrode using dialysis membrane. The estimation of glucose concentration was performed by measuring the reduction in the dissolved oxygen concentration. The idea proposed was industrialized by Yellow Springs Instrument Company (Ohio, USA) in 1975 and the glucose sensing analyzer, based on the amperometric detection method, was successfully relaunched. Later, the detection of urea concentration using direct potentiometric method was reported (Katz and Rechnitz 1963). Further development in the sensor was made with the introduction of urea electrode (Guilbault and Montalvo Jr 1969). The biosensor based on microbe technology was pioneered in 1976 (Cammann 1977) and the term "biosensor" was introduced. The technology was further developed with the application of artificial redox mediators (Kulys and Svirmickas 1980) and later, the first ferrocene-mediated amperometric glucose biosensor was proposed (Cass et al. 1984), which was industrialized by Medi Sense Inc. in 1987. As a continuation of the work done by various researchers, IUPAC introduced the definition of biosensors analogous to chemosensors as "Biosensor is a device that uses specific biochemical reactions mediated by isolated enzymes, immune-systems, tissues, organelles or whole cells to detect chemical compounds usually by electrical, thermal or optical signals" (Monosik et al. 2012). Followed by this great achievement, a number of research communities from various fields like very-large-scale integration (VLSI), life science, physics, chemistry, and material science initiated their interest for the development of biosensing devices incorporating more sophisticated, reliable, and miniaturized biosensing capabilities. In 2007, an implanted glucose biosensor (freestyle navigator system) was operated for 5 days and the work on the development of biosensors continued its advancement exploiting the enzymes, antibodies, and microbes in combination with various types of transducers (Weinstein et al. 2007).

The term "biosensor" signifies a device which is fabricated by the combination of two parts: bio-recognition element and the sensor element. In the biosensors, bio-recognition elements like enzyme and antigen recognize the selective analyte to be detected, while the sensor element is used for the transduction of the change in biomolecule concentration into an electrical/optical signal. Hence, the sensor element is also recognized as the transducer. The bio-recognition element should be highly specific to the analyte, to which it is sensitive for the change in the electrical/optical signal. Biosensors are used for industrial applications, biomedical diagnostics, food safety, and military applications. In the beginning, major application of biosensors was the sensing of glucose concentration in blood due to its plentiful application in biomedical industry. However, nowadays, biosensors are used in a number of commercialized applications such as fabrication of devices for the calibration of foods and beverages concerning the human health, development of the devices for environmental sampling, and noninvasive instruments for clinical analysis (Mello and Kubota 2002, Patel 2002). However, the industrialization of biosensor devices is quite slow due to various technological issues and limitations. For example, it is necessary to prevent the biosensor from contamination and is challenging too at the same time, because of the presence of semiconductor materials along with the biomolecules in biosensing devices.

1.2.2 Gas Sensors

The development of the gas sensors became the topic of discussion after the discovery of the effect of harmful gases on human health. During the nineteenth and early twentieth centuries, coal miners used to take canaries to the coal tunnels for the early detection of life-threatening gases such as carbon monoxide, carbon dioxide, and methane. Canary, a very songful bird, stops singing and eventually dies in the continuous presence of gases. This was used as a signal for the detection of gases (Thompson 2004). The first gas detection and monitoring system was a flame safety lamp invented by Darway in 1815. In the lamp, the height of the oil flame was adjusted to a specific position in fresh air which was kept inside the glass sleeve with a flame arrester. High flame was the indicator of the presence of methane gas, while the low flame was for low oxygen. Later, during 1926–1927, Oliver Johnson introduced a gas sensor based on catalytic combustion method. After this, various types of gas sensors based on methods such as electrochemical and catalytic combustion were commercialized. In the 1970s, further establishment of various types of gas sensors was carried out with different techniques including humidity analysis, semiconductor combustibility, and solid electrolyte oxygen. Later, extensive attempts were made not only for the advancement of these sensors but also for the implementation of several types of new gas sensors, which were in great demand for various applications such as environmental reservation, energy saving, health safety, air pollution monitoring, etc. Gas sensors are the most important part of environmental monitoring system because of the gases produced from various sources such as home appliances (chimneys to air conditioners) and various industries that expel toxic waste products. Hence, continuous monitoring of environment is required for human safety and unpolluted atmosphere. Due to these reasons, the gas-sensors-producing industries have been developed up to a significant level to meet the requirements of providing a standard and clean atmosphere for better living. However, further improvement in the gas-sensing technology is required for the improvement of the quality of life and sustainability of our society.

1.2.3 Chemical Sensors

Chemical sensor is used for the determination of chemical composition of the surrounding medium which can be either in gaseous phase or in liquid phase. If the target to be detected is in the gaseous phase, then the device prepared for the detection of gases is generally termed as the gas sensor. The sensor recognizes the presence of the target species and transforms it into a physical signal. Thus, a variation in the characteristic physical signal is observed due to the variation in the concentration of target species. Being a broad area, the technique of gas sensing is now considered as a separate field. The first significant work for the development of chemical sensor was reported by Arnold Backman in 1932 with the invention of modern glass electrode for sensing application, which was later commercialized. Later, in 1936, he designed and patented a pH meter based on a pair of glass electrodes. In the patented device, the electrode pair was suspended in the test solution and the potential difference between the electrodes was used to measure the pH value of the test solution. In 1937, the application of solid-state electrodes such as silver halide and fluoride-specific electrodes for chemical sensing was introduced (Kolthoff and Sanders 1937). The chemical sensors utilizing antibodies as the recognition system for the detection of biological samples were developed (Moore and Pressman 1964). Almost 2 years later, the selective and reversible properties of alkali metal ions were reported (Stefanac and Simon 1966). This study was further used for the development of K^+-selective electrochemical sensor (Frant 1994). An electrochemical serum analyzer for NASA's Space Shuttle was proposed and the first industrialized sodium/potassium analyzer in blood sample was reported, which was named as SS-30.

Followed by the development of chemical sensors based on naturally occurring antibodies with highly specific properties, the researchers started focusing on the fabrication of artificial ligands for the detection of various analytes. The studies for the development of artificial receptors were pioneered by Pederson, Lehn, and Cram who were awarded the

Nobel Prize in 1987 (Pedersen 1967, Cram 1988, Lehn 1988). Various synthetic ligands such as synthetic macrocyclic polyethers, crown compounds, cyclophanes, and macrohetero-bicyclic were developed for the chemical sensor applications by the team. In the same period, an optical chemical sensor was introduced by evaluating the first optical potassium test based on dry reagent chemistry (Keller 2008). The method was commercialized by Ames Division. In 1968, the first fluorosensor for oxygen detection was used for the biomedical applications (Lubbers and Optiz 1975). Lubbers suggested the name of optical sensor as, "optrode" analogous to the "electrode," which was further corrected as "optode." A considerable amount of work was carried out by various research groups in 1980s using optical chemical sensors for food safety and environmental monitoring. Later in 1991, after the remarkable growth of the chemical sensor industry, IUPAC gave the definition of the chemical sensor as, "A chemical sensor is a device that transforms chemical information, ranging from concentration of specific sample component to total composition analysis, into an analytically useful signal" (Hulanicki et al. 1991). Few more definitions of the chemical sensors were reported in a more general way by various authors. A well-known one by Wolfbeis is as, "Chemical sensors are small sized devices comprising a recognition element, a transduction element and a signal processor capable of continuously and reversibly reporting a chemical concentration" (Wolfbeis 1991).

1.3 Sensor Components

As mentioned above, the sensor can be divided into four components: analyte, recognition unit, transducer, and analyzer (Diamond 1998). The working of the sensor is based on the interaction of analyte to be sensed with the recognition unit, which makes a significant change in the characteristics of the input signal that is detected with the help of a transducer, by a detection unit or analyzer. The analyte/sample may be biological (such as glucose, cholesterol, proteins, etc.), chemical (such as pH, metal ions, etc.), and environmental (gases like CO_2, CH_4, etc.). Recognition unit is the most important component of the sensor. The analyte to be detected reacts with the recognition unit through a predesigned interaction. Numerous types of natural and artificial receptors have been introduced by researchers for developing the sensing devices. These will be discussed in further sections. Transducer is the next salient component of the sensing unit. It is used for the conversion of one type of signal into other form, thus converting the analyte concentration into corresponding signal. Analyzer or detector is the last but essential part of the sensing unit. The analyzer is required for detecting/measuring several types of signals which correspond to recognition of the analyte. A block

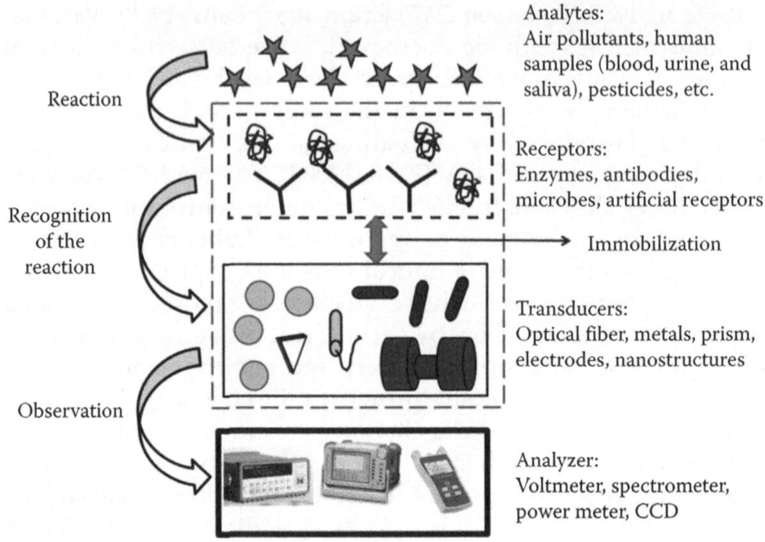

FIGURE 1.2
Block diagram of the sensor components, showing the analyte/s that needs to be interacted with recognition element/s interfaced with the transducer and then the results of interaction are transferred to the detection and analysis unit.

diagram showing the sensor components with possibilities in their selection and path in which the working proceeds is shown in Figure 1.2.

1.3.1 Analyte

Analytes (samples) are various types of atmospheric, chemical, biological elements which need to be detected. The analyte may be either in liquid state or in gaseous form. According to the type of analyte, sensors can also be classified as chemical, gas, or biosensors. Various hazardous gases, atmospheric pollutants, chemical wastes through industries, biological elements present in blood/urine, nutrients and vitamins, pesticides, etc. are the few examples of analytes needed to be sensed. Several types of ambient conditions and physical parameters such as pH, temperature, and pressure also need to be sensed as they play important roles in the industries for the fabrication of various medicines, devices and need to be monitored for the human and environmental safety. Thus, these can also be considered as analytes. Most common gases such as hydrogen, hydrogen sulfide, chlorine, methane, carbon monoxide, water vapor (humidity), ethanol vapor, etc. can be treated as analytes in the gas-sensing unit. The concentration of metal ions in water should be optimized because higher concentration of various metals such as lead, calcium, cadmium, iron, arsenic can result in unhealthy conditions. The vitamins A, B, C, D, E, and K are essential for human body which can be found in various food stuffs such as egg, milk, fish, etc. except vitamin D, which is obtained from direct sunlight. Each vitamin has its own functionality and its deficiency causes many health-related problems in human body. Thus, their concentration/availability should be sensed in various foodstuffs as well as in blood. In blood, the analysis of various biological elements for clinical applications is possible. Other analytes such as pesticides, minerals, antibodies, and antibiotics are required to be detected for a healthy life. The concentration of analyte is recognized by its interaction with the receptor/recognition unit.

1.3.2 Recognition Unit

Recognition unit, generally termed as the receptor, is the most important part of any sensing device since it decides the interaction nature of the sensor with the analyte and hence the result. This is also called as the "heart of a sensor." The recognition unit or receptor should be selective and specific, while interacting with the target analyte to be sensed. It should interact with the analyte reversibly, and specifically without any interference from other similar chemical analytes except the target analyte. Generally, in sensors, receptors are immobilized over the surface of the transducer, so that transducer can transform the interaction between analyte and recognition unit into a readable output signal which can be analyzed by the analyzer. For example, in the case of enzymatic biosensors, the enzyme which interacts with a specific analyte is the receptor of the sensor. In the case of few physical sensors like temperature and pressure sensors, no receptor is required as the working of the sensor is based on the physical properties of materials used in the fabrication of the fiber/transducer. A detailed discussion on the types of receptors and their compatibility with the transducer is given in the next section.

1.3.3 Transducer

By definition, transducer is a device which converts one kind of signal into another form. In the case of sensors, transducer is the unit which helps to measure the change in the properties of the receptor due to its interaction with the analyte, by converting the changes in to some readable signals. In optical sensors, the readable signal is the change in the parameters of the output light collected by a detector/analyzer. The phenomenon by which this conversion is performed is the transducing phenomenon which can be optical, electrical, etc. Further, the element which realizes the transducing phenomenon is called as the transducer such as metal films, nanoparticles, fluorophores, etc. Generally, the recognition unit is immobilized over the transducer surface which is sensitive toward the changes over its surface properties. When the analyte interacts with the receptor unit, the optical properties of the receptor unit change. Since the receptor is immobilized over the transducer surface, the properties of transducer surface change and due to the transducing mechanism (such as surface plasmon resonance [SPR]), the change in the specific parameter (such as minimum transmission at a specific wavelength, resonance wavelength) is obtained at the detection unit. Thus, the light from the source interacts with the optical transducer unit and is collected at the output after undergoing changes in its parameters. The change in the parameter value specifies the interaction and sensitiveness of the transducer surface to the analyte.

1.3.4 Detector/Analyzer

After the working of the transducer, the change in the signal is measured by the detector. These are selected according to the transducer used, for example, thermometer for measuring the temperature, ammeter for the measurement of current, voltmeter for the voltage observation, magnetometer for analyzing the magnetic field, optical power meter for measuring the light intensity, and so on (Gupta 2006). There are various types of optical detectors such as spectrometer, optical power meter, photodetector, charge-coupled device (CCD), fluorometer, etc. Since the present book deals with the optical sensors, we shall now provide a brief introduction of some of the optical detectors used in the optical sensors.

1.3.4.1 Spectrometer

Spectrometer is one of the most broadly used detectors in optical sensors. It is used for the measurement of light intensity for a particular wavelength range. The spectrometer consists of an entrance slit, collimator, dispersing element (prism or grating), focusing lens, exit slit, and a detector. The grating-based spectrometer has resolution better than the prism-based spectrometer. The grating with few hundred lines per mm is placed at a specified blaze angle, which decides the operating wavelength range of the spectrometer. For the detector of the spectrometer, usually a CCD linear array with thousands of pixels is used. These are responsible for the fast processing of the data. The selection of all the components of a spectrometer such as entrance slit, diffraction grating, order shorting filter, and the detector coating strongly affects the performance parameters of the spectrometer such as spectral range, resolution, sensitivity, and stray light.

1.3.4.2 Optical Power Meter

Optical power meter is used for the measurement of the power in an optical signal. It is usually used for the determination of the power of the light guided through fiber-optic cable. It is made up of a calibrated sensor interfaced with an amplifier and display. It is also used for the observation of power loss in the optical signal when light is guided through the fiber. In the case of sensing applications, it is used for analyzing the output power through the fiber-optic sensor after transducer recognizes the interaction of the analyte and the recognition unit (Gupta 2006).

1.3.4.3 Charge-Coupled Device

A CCD is a light-sensitive integrated circuit which is used to store and display the data for an image. The working of CCDs involves the conversion of each pixel (picture element) in the image to an electrical charge, the intensity of which is related to the color in the color spectrum. These are used as sensors in digital cameras and video cameras for the recording and storage of steady and moving images. The CCD captures light and transforms it to digital data, which is stored by the camera. The quality of an image captured by a CCD depends on the resolution of the sensor. In digital cameras, the resolution is measured in megapixels (or thousands of pixels). Therefore, an 8MP digital camera can capture twice as much information as a 4MP camera. The result is a larger photo with more detail. These are used in the biosensors for the human-noninvasive monitoring under clinical situations.

1.4 Sensor Performance Parameters

In this section, we shall discuss various static and dynamic parameters which are used for the characterization of the sensor to evaluate its performance. On the basis of these parameters, the sensor performance can be enhanced which is greatly required for the advancement in the sensor technology. These are the following parameters:

- *Sensitivity*: Sensitivity is one of the most important parameter of the sensor. It can be measured by calculating the change in the sensor output signal with respect to unit change in analyte concentration which is to be detected. It can also be evaluated from the slope of the calibration curve, which is generally the plot between the measured output signal and analyte concentration.
- *Selectivity/specificity*: Selectivity of the sensor corresponds to whether the sensor responds either to a group of analytes or the sensor responds specifically to a single analyte. The sensor is highly selective, if it responds only to the target analyte to be sensed in the presence of interfering analytes.
- *Operating range*: The analyte concentration range to which the sensor operates is the most important parameter and it decides the sensor's applicability in the industry/hospitals, etc. Usually the sensors'

operating ranges are fixed because an increase in the sensor operating range decreases the sensitivity of the sensor.

- *Linear range*: The linearity range of the sensor defines the sensor's operation with fixed sensitivity, which makes the sensor to be operated easily. This can be extracted from the calibration curve of the sensor by fitting an ideal straight line. In other words, this tells us the analyte concentration range up to which the output signal varies linearly.
- *Limit of detection*: The lowest analyte concentration which can be detected by the sensor is defined as the limit of detection (LOD) of the sensor. The evaluation of the value of LOD depends upon the type of sensor fabrication and the detection system.
- *Limit of quantification*: Limit of quantification (LOQ) is also referred to as the lowest analyte concentration which can be detected by the sensor, considering the standard deviation in the output signal.
- *Resolution*: It is the lowest change in analyte concentration which can be recognized by the sensing device. This is an important parameter for the sensors in flowing streams.
- *Response time*: Response time of a sensor denotes the time taken by the sensor to show a saturated output for a given change in the analyte concentration. Generally, it is defined as the time required by the sensor to respond for the analyte detection.
- *Repeatability and stability*: As named, these terms correspond to the sensor's ability for reproducing its output response after characterizing the sensor up to certain cycles and maintaining its performance for a specific time period. These parameters play the most important role for the sensor to be commercialized.
- *Hysteresis*: It is the main characteristic of the sensor which decides the difference in sensor's response for an increasing and decreasing analyte concentration. It is normally given as the percentage change in the output response after the interaction and removal of the analyte from the recognition unit of the sensor. Zero percentage hysteresis refers to the exactly reversible nature of the sample.
- *Shelf life*: This parameter refers to the time period for which the sensor will operate. If the sensor is not usable for a favorable time period, then it affects the sensor's preference.

1.5 Biosensor Classifications

An independent sensor incorporating a biological material in its active form in the system, to interact with the analyte which is chemical/biological, is generally considered as a biosensor. The design, selection, and application of biosensors are critical since its interaction with the desired analyte to be sensed is always affected by interferands. The application needs to be performed in external environments, harsh atmospheres, laboratories and health clinics, and even for *in vivo* and *in vitro* measurements in human body. The analytes and the active biological elements may also need to be sterilized in some cases, which consume time, high-cost equipments, man power, and control measurements. However, the possibility of purifying these increases the interaction of sensor with the analyte and provides better analysis.

The necessity of sensors in the area of health, environment, pharmaceutical, agricultural, water monitoring and food quality, hygiene, drug and medicine development, and dialectic cases is increasing. The sensor should be easily available, economical, portable, easy to handle, commercialized into a small module so that even laymen can use it for agricultural applications, food safety, and environmental monitoring. Thus, the processing and the analysis should be performed in a smart way. In the case of biosensors, especially, the analysis should be rugged to make the sensor extremely selective, sensitive, reliable, recoverable, and vigorous. This is where we can find the difference between biosensors and bioanalytical schemes, which require multiple processing and are of onetime use. For the use of biosensor as a tool for diagnosis, in medical field such as cancer/tumor cell determination, the sensor needs to be reproducible with capabilities of multianalyte detection (Tothill 2009).

Biosensor works either in direct or indirect monitoring mode, depending on the kind of reaction or analysis the sensor is intended to perform. In direct monitoring mode, the analysis is done by monitoring the direct products and reactions occurred due to the interaction of analyte (Gauglitz 2005), whereas in the case of indirect monitoring mode, the analysis is performed indirectly through the inhibitors or catalysts consumed for the reactions. The central unit of every biosensor is the bio-recognition element synced with the transducer unit. The functioning of the biosensor depends on its recognition element and transducer. Hence, biosensors are mainly classified on the basis of their recognition element and transducer units. In the following sections, we shall discuss the classification of biosensors and their importance for different applications and measurements by correlating the recognition and transducer modes.

1.5.1 Classification Based on Transducer

The transducer part of the biosensor system is the unit, which provides bidirectional signal conduct. The transducer part determines the signal mode available at the output for analysis and readable by humans. There is a predetermined relationship between the output and the input signal in the biosensor system, and by calibrating the results from the analyzer one can evaluate the measurements and changes. The transducer element in a biosensor can be chosen according to the design and substrate of the sensor probe. Thus, based on the transducer element, biosensors can be broadly classified into five main types as electrochemical, optical, calorimetric, mass-sensitive, and light-addressable potentiometric transducer (Patel 2002, Reyes et al. 2003).

1.5.1.1 Electrochemical Transducer

Electrochemical types of transducers are the commonly used ones because of their fast response, automation possibility, and compatibility with modern technologies of fabrication, portability, and minimal requirement of power (Thevenot et al. 2001). These sensors are usually used in point-of-care applications such as glucose sensor. In the case of electrochemical sensor, the substrate is the electrode, which is chemically modified. This can be done by modifying the electrode with ionic/conducting/semiconducting material along with a chemical/biofilm (Durst et al. 1997). There are two electrodes, one a working electrode which performs sensing and another a reference electrode dipped in the electrolyte. Electrochemical sensors can be potentiometric, amperometric, conductimetric, and impedimetric.

An amperometric electrochemical sensor setup consists of working electrodes (cathode and anode) and a reference electrode dipped in an electrolyte as shown in Figure 1.3. For a particular potential applied between the electrodes, a chemical reaction occurs due to the oxidation or reduction of electroactive species. The process of oxidation or reduction generates current in correlation with the concentration gradient of the electroactive species/analyte which is measured by an ammeter. For the sensing purpose, the receptor is immobilized over one of the electrodes. When analyte is added into electrolyte, the current starts flowing with the electroactive species toward the electrodes. In the presence of receptor, the analyte binds with the receptor and the conductivity of the receptor unit changes. Since the receptor is immobilized over the electrode, the conductive properties of the transducer surface change. This results in the change in current flowing through the electrode setup, which is analyzed by the ammeter. Thus, by observing the change in current, before and after adding the analyte, in ammeter, one can detect the presence of analyte. Electrodes are generally made up of metals/steel/carbon that are conductive in nature and are modified according to the application. The bio-recognition element and the immobilization methods with matrix can be varied accordingly, which may also use an electron mediator or a selective/discriminative membrane. Amperometric sensors are fast, more sensitive, and exact than other types in electrochemical system. The calibration curve of these sensors is, in general, linear. The first biosensor using amperometric transducer was designed for glucose monitoring based on the consumption of oxygen (Updike and Hicks 1967). This kind of amperometric sensor, which performs analysis based on the consumption of indicator or formation of

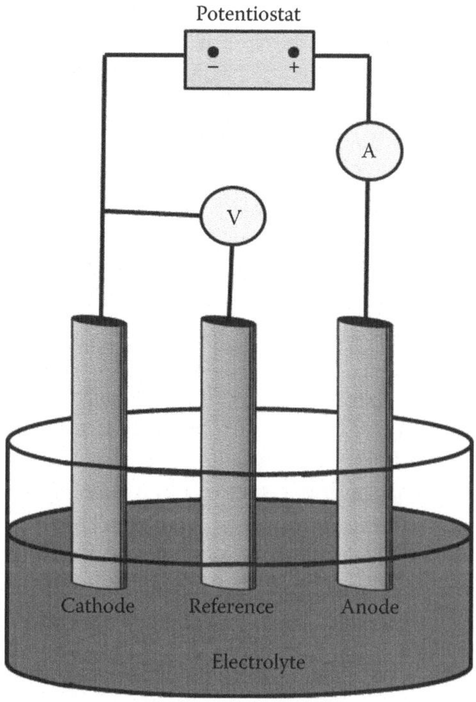

FIGURE 1.3
Schematic of a general amperometric transducer setup.

product, comes under first-generation sensors. The sensors utilizing electron mediators and performing redox reactions come under second generation, whereas the reagent-less sensors come under third generation. The third generation uses enzyme or antigen–antibody coupling over electrode which transforms the electrode by the catalytic reaction and generates a catalytic current for the analysis. The amperometric-electrochemical sensor holds applications in food industry for the detection of bacteria, food pathogens, and food adulterations.

A potentiometric transducer measures the potential drop between the reference and working electrode as the signal to be analyzed or the potential generated across an ion-selective/semipermeable membrane which separates two solutions, when no current is flowed (Frant 1994). The potentiometric transducer can be based on field-effect transistor (FET), where an ion-selective membrane replaces the gate of FET to give ion-sensitive FET (ISFET) or enzyme immobilized to give enzyme-sensitized FET (ENFET). Thus, the transducer element can be either an electrode with ion-selective membrane (ion-sensitive electrode) or a membrane which works as bio-recognition element. Potentiometric-electrochemical sensors are commonly used as pH sensors or gas sensors and the potential difference is in proportion to the logarithm of the gradient in analyte concentration. A disadvantage of the potentiometric-transducer-based sensor is the affinity of the system to the buffer competency.

Another kind of transducer under electrochemical type is conductimetric-electrochemical transducer. In this type, the analysis is carried out by measuring the electrical conductivity which changes during biochemical reactions that release electrons. Usage of this kind of electrochemical sensor is slightly restricted in biosensing applications due to its implementation difficulties especially when the bio-recognition element is an enzyme.

The last class of electrochemical transducer, came up in 1992, is impedimetric which measures the impedance or resistivity of the cell. This is also called as conductometric since the resistance and conductivity are inversely proportional. The main drawback of conductometric-electrochemical sensors is the poor sensitivity of the system. Both conductimetric and conductometric sensors are used for food quality monitoring, air and food sanitation microbiology. Portable impedance-electrochemical biosensor for microorganism detection is available in markets, in which the analysis is performed by measuring the impedance/conductance caused by the metabolic reaction that increases the conductance implying a decrease in impedance.

1.5.1.2 Optical Transducer

The advancement in optical techniques and their implementation using substrate from prism to grating, waveguide, and finally the fiber has made the evolution of optical biosensor as a principal real-time analytical tool. The optical biosensors available are based on a large number of techniques such as reflection, refraction, interference, diffraction, polarization, absorption, fluorescence/phosphorescence, chemi/bioluminescence, scattering, spectroscopy, etc. (Santos and Farahi 2014). High sensitivity, small response time, robustness, simplicity, and small size apart from immunity to electromagnetic interference make the optical sensor easy to handle and install at various locations. If the substrate used for optical sensor is an optical fiber, then the possibility of remote sensing and capability of online detection of analyte become the additional advantages.

The emission of energy in the form of radiation when a molecule/electron excited by a photon of high energy moves from a higher energy to lower stable energy level is termed as photoluminescence. When the excitation of molecules occurs by some chemical reaction, then the process is called as chemiluminescence, whereas in bioluminescence, the biological or biochemical reaction/metabolism occurring in organisms results in the excitation of molecules. Thus, in chemiluminescence, light source is not required. The chemi/bioluminescence depends on the type of molecule, concentration, temperature, and environmental conditions. Luminescence-based optical sensors are used for the detection of carbohydrates, alcohol, etc., where the fermentation process needs to be controlled. A considerably selective and sensitive sensor can be designed using fluorescence technique where the frequency variation in the emitted electromagnetic radiation is analyzed, implying that the analyte absorbs one wavelength/frequency and emits another (Wang et al. 2002). Fluorescence occurs because of the absorption of the input radiation and the generation of radiation of lower frequency after interaction of the input radiation with the fluorophore. The technique of fluorescence emission is used for the analysis of contaminants in food, pharmaceuticals, etc. The concentration of the analyte is determined directly or indirectly from the duration/intensity of the fluorescence emission. The emission and excitation occur at particular levels of energy and hence there is a special spectrum of fluorescence for each analyte and for each molecule of analyte. This makes the technique selective for a particular sensing application. Fluorescence is a kind of photoluminescence, where the transition occurs between the "excited singlet state of lowest energy and the ground singlet state" by the general definition. Phosphorescence is another kind of photoluminescence in which the electron changes its spin during the transition since it moves in the same level. The mostly used optrodes (optical + electrode) for biological application are based on fluorescence phenomenon. Generally, the device consists of light source, detector, and sensing probe made up fiber. The sensing fiber head is coated with the bio-recognition element. While choosing the technique of fluorescence for sensing or analysis of analyte, the parameters that help in diagnosis are the intensity and lifetime of fluorescence emission. The intensity measures the analyte concentration, whereas the lifetime shows the period for which the electron/fluorophore remains in the excited state. The lifetime is measured by calculating the shift in phase between the emitted and the excitation light since lifetime and phase shift are directly proportional. Fluorescence-based optical biosensor is commonly used as immunosensor- or catalyst-mediated sensor such as enzyme-immobilized ones. The emission of fluorescence is proportional to the analyte concentration in the case of immunosensor, while the catalytic reaction decides the emission in the catalyst-mediated sensor.

Scattering-based optical transducer uses either forward scattering or backward scattering mode, especially in the case of fiber-based system. Both ends of the fiber are operated in the case of forward scattering, while only one end of fiber is operated in backward scattering mode. Although the mode of back scattering provides advantages over forward scattering in terms of high sensitivity and easy operation, the method also holds disadvantages such as large response time, low-intensity signal, etc. This difficulty is usually overcome by integrating the signals received over a number of times. Scattering mechanism can be elastic or inelastic depending on the wavelengths or frequencies of the incident and scattered light. If the incident and scattered light frequencies/wavelengths are same, then it is called elastic scattering, for example, Rayleigh; while the frequencies/wavelengths are different, then it is called inelastic scattering, for example, Raman and Brillouin scattering. Generally, the scattering-based techniques are employed using interferometric methods. Rayleigh scattering occurs due to the change in the dielectric constant, which is caused by the variation in the composition and density as the light traverses through the optical fiber. Rayleigh scattering-based technique is in use for the environmental monitoring such as identification of explosives. The technique can be used in a better way by enhancing the surface using the Raman-active nanostructures or molecules adsorbed on the prepared surface of a metal, to be called as surface-enhanced Raman scattering. The enhancement is carried out either chemically or electromagnetically. The technique has applications in protein detection and biomedical sensing with advantages of limited response time, trace element identification, which make it useful in the fields of pharmaceuticals, food industry, and environmental monitoring (Han et al. 2009). Similarly, diffraction-based optical transducer-employed sensors are in use. Such sensors, in general, make use of grating structures that offer multiplexing potential and compatibility to work with fiber technology which helps in biosensing applications.

Absorption-based optical sensors, in general, use the amount of light absorbed by the analyte to quantify its variation in concentration. The analysis is performed by calibrating the transmitted light, after absorption by the medium, with the incident input light. The transmitted light intensity from an absorbing medium, ideally considering uniform absorption, is, mathematically, related by the Beer–Lambert law as

$$I = I_0 \exp(-\varepsilon C \Delta l),$$

where I_0 and I are the intensities of the incident and transmitted light, ε is the extinction coefficient of the absorbing analyte, C is the concentration of the analyte that absorbs light, and Δl is the thickness of the analyte medium. The constant, εC, is called the linear attenuation coefficient which can be equated to the absorption coefficient if the scattering of light by the medium is zero. The thickness/depth of the medium and the transmitted light intensity are inversely proportional to each other. The absorption of light by the analyte depends on the wavelength of the incident light, which holds a major application in biosensors. Although the method is cost-effective and simple, its application on biological samples is limited because of limitations of samples/biological analytes to absorb light, which causes a large background signal in the spectrum. Another type of optical-based transducer employs evanescent wave absorption and is generally used in environmental monitoring. This kind of sensor is generally realized on optical fiber with the possibilities of modifications in fiber structure, fiber geometry such as U-shaped, and tapered to enhance the sensor's performance. At the interface of core-cladding/medium where the total internal reflection occurs, evanescent wave is generated by the emanation of electromagnetic energy with a decaying field in the medium. The sensor system is operated in intensity modulation scheme and the change in the imaginary part of the refractive index of the medium around the core causes the change in the intensity of the transmitted light. A simple evanescent wave-based optical transducer system has the physical/chemical/biological measurable on the unclad region of optical fiber. This unclad region of the sensor probe is called the sensing region. Immunosensors are possible to realize with evanescent wave absorption-based optical sensors. The evanescent wave has an average penetration depth of around 200 nm. A modified type of evanescent wave absorption-based optical transducer uses fluorescence along with it, where the antigen–antibody complex on the interface absorbs evanescent wave and emits the fluorescence. The concentration of analyte can be determined from the emission spectra. Evanescent wave absorption-based optical sensor system does not require any coupling components and provides sensors with better sensitivity compared to sensors using bulk optics.

Another category of optical transducer is based on the refractive index. Although grating, interferometric, Raman scattering, and evanescent wave-based transducers can also work with the change in the refractive index, the most importunate techniques that come under this category are the SPR, localized surface plasmon resonance (LSPR), and lossy mode resonance (LMR). The calibration curve of these kinds of refractive-index-based optical sensors shows the change in the sensor calibration parameter with the change in the refractive index, thereby measuring the analyte concentration indirectly. The technique can be realized on prism, optical fiber, or grating. The phenomenon of SPR accomplished on optical fiber has been a great achievement since 1993 when prism was replaced by the core of the fiber for SPR-based sensor (Jorgenson and Yee 1993). The combination of surface plasmon technology and optical fiber has brought advantages such as miniaturization of the probe with real-time implementation, online monitoring and remote sensing facility, simple, easy-to-handle, and cost-effective probe. This combination has overruled all the existing techniques. The usage of metal nanoparticles/nanostructures instead of metal film has introduced LSPR. The combination with optical fiber brings biosensor with high sensitivity and better performance. The basics and developments of SPR and LSPR phenomena will be discussed in detail in Chapter 2.

A latest sensing technique, employing the optical transducer based on refractive index, is LMR where a metal oxide/suitable polymer/ceramic replaces the metal in SPR technique. The method has been reported to have huge sensitivity even for biosensing. The method requires more studies and also developments to implement it for biomedical applications that would be a huge turnover in the medical field. A combination of SPR/LSPR and LMR has also been explored in the literature. The methods as well as their combinations provide high-performance biosensors with good biocompatibility and long life if implemented using optical fiber substrate.

1.5.1.3 Mass-Sensitive Transducer

In a mass-sensitive transducer employed in biosensor, a bio-layer is kept in contact with a piezoelectric element and the change in mass, density, dielectric, electric, or elastic properties of the bio-layer is measured. The mass-sensitive transducer can be divided into acoustic wave- and cantilever-based transducer. In acoustic type, bulk acoustic wave and surface acoustic wave propagation transducers are employed. Further, the operation of sensor is based on the resonance of the oscillating crystal (usually quartz) at its basic frequency. The oscillator part is the one with the quartz crystal connected to an amplifier. The resonance frequency changes according to the properties of the bio-layer connected. In the case of surface acoustic wave propagation transducer, the bio-recognition layer is placed over the substrate with input and output ports. The wave traverses along the surface of the bio-recognition layer and the properties of wave are affected by the properties of the bio-layer. Using an electrical signal of particular frequency, the piezoelectric crystal is made to vibrate. The interaction of the analyte with the bio-recognition layer changes the mass of the crystal which changes the resonance frequency. The materials that exhibit piezoelectric properties are quartz, tantalite, tourmaline, zinc oxide, etc. Molecular sensing is an important application of piezoelectric sensors with label-free detection possibilities.

The second type of mass-sensitive transducer is based on the microcantilever. The cantilever-based-transducer-type sensor generally measures the change in vibration frequency or bending of cantilever. The interaction of the analyte with the microcantilever surface causes a tilt in the cantilever. The method has advantages of real-time detection, usage of small quantities of analytes and bio-layer elements, good LOD, etc. Microcantilever-based sensors are reported for the implementation of cancer biomarkers.

1.5.1.4 Calorimetric Transducer

The sensor monitors the change in energy, especially, in the form of heat. The bio-recognition layer is immobilized on thermistors or temperature sensors (Hoffmann et al. 2007). The interaction of analyte on the bio-recognition layer results in heat generation, which is sensed by the sensor. The heat absorbed or produced depends on whether the reaction is exothermic or endothermic, and is related to the total number of molecules involved and molar enthalpy. The thermistor is independent of the optical or chemical properties of the analyte. Depending upon how heat is transferred, the calorimeters are divided into three main categories: isothermal, heat conducting, and isoperibol. In isothermal, the temperature of the reaction chamber is maintained constant by Joule heating and the quantity of energy that is necessary for that is measured. In the case of heat conduction calorimeter, there is an isothermal heat sink around the reaction chamber and the heat difference between the two is measured (Maskow et al. 2012). In isoperibol, the system is the same as that of heat conduction one, but the reaction chamber is thermally insulated. Calorimetric biosensors are mainly used in food industry and pharmaceutical investigations.

1.5.1.5 Light-Addressable Potentiometric Transducer

The light-addressable potentiometric transducer is a combination of optical and potentiometric/electrochemical transducer systems. The method was first time introduced in 1988 (Hafeman et al. 1988). It is a semiconductor device having heterostructure and generates photocurrent when excited by light of enough intensity. The amplitude of the photocurrent generated is proportional to the change in surface potential. Thus, a potential variation caused by an electrochemical means is measured in terms of optical parameter. The advantages of this method include easy fabrication and encapsulation. Since it is an integrated sensor system by itself, it has achieved a level of miniaturization with capability of array sensing.

1.5.2 Classification Based on Bio-Receptors

The bio-recognition element in a biosensor interacts with a specific analyte in a sensor. The bio-recognition elements are broadly based on catalytic reaction, affinity, and hybrid-complexing. In catalytic-reaction-based sensors, the catalytic/chemical reaction occurs when analyte comes in the vicinity of the bio-recognition element. In the case of affinity-based sensor, there are specific interactions between the analyte and the receptor/bio-recognition element. The affinity-based sensors are inherently selective in nature. In hybrid-complexing-based sensors, bio-recognition elements undergo complex reactions and hybridizations while sensing. The general bio-recognition elements include enzymes, nucleic acids/aptamers, antibodies, cells, microbes or microorganisms, membranes, supramolecular structures of cell/cell organelles, tissues, organic molecules, and artificial receptors. Figure 1.4a shows the bio-recognition elements with examples of their applications toward environmental and biological analytes, while Figure 1.4b shows the possibilities of transducer elements that are commonly chosen for each bio-recognition element.

Enzymes are the most common class of bio-recognition elements used in sensing. The first biosensor reported utilized glucose oxidase that was entrapped at a Clark oxygen electrode (Clark and Lyons 1962). The enzymatic reactions are specific, which produce predetermined products where the enzyme acts as a catalyst by deciding the rate of generation of intermediate and by-products. Although the enzymatic reactions are certain to occur, the reaction and its rate are influenced by various factors such as enzyme mounting, stability, pH, temperature, and the environmental conditions in which the sensor is supposed to work. Sometimes, coenzymes are also

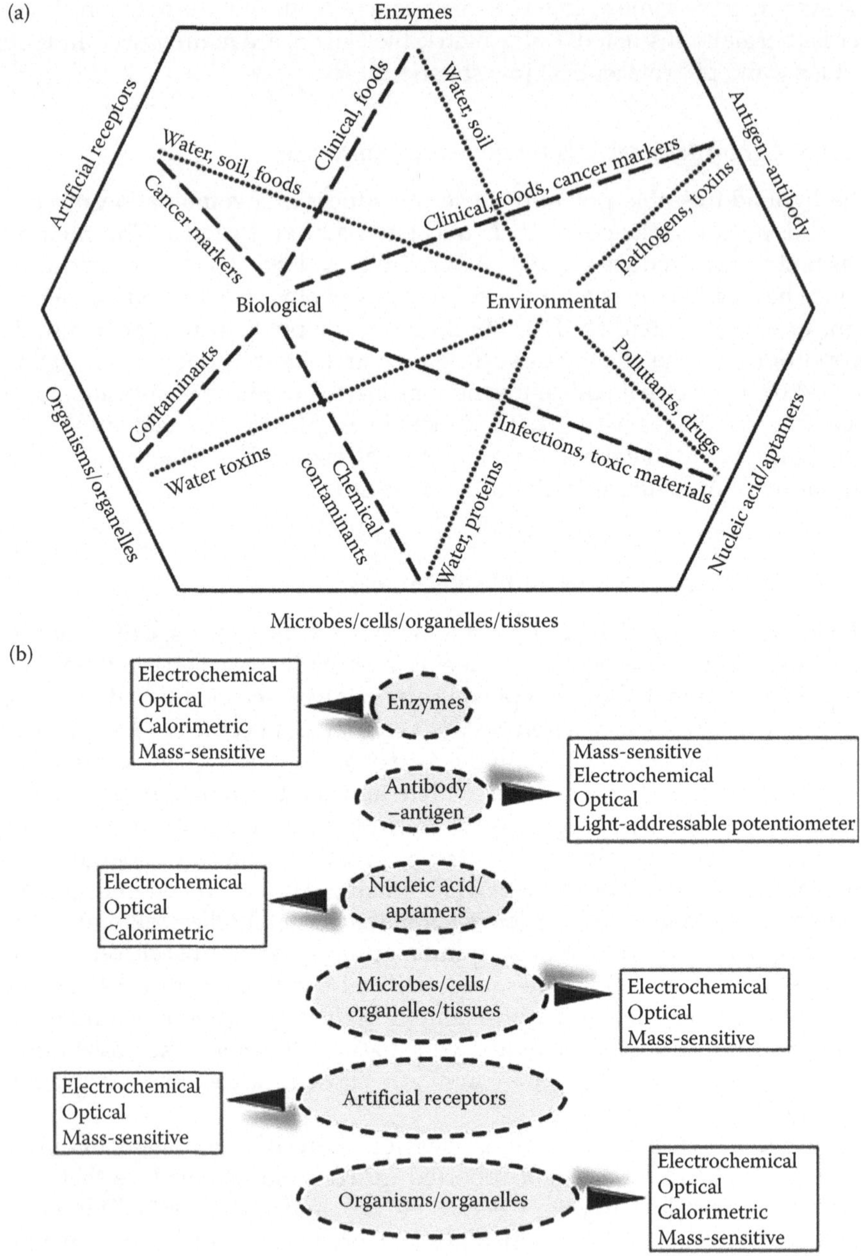

FIGURE 1.4
(a) The groups of bio-recognition elements with their common applications in biomedical applications and environmental monitoring, (b) transducer elements that are interfaced in common for environmental and biomedical applications for the groups of bio-recognition elements.

used to ensure the stability and activity of the enzyme. There are different categories of enzymes according to the reaction catalyzed such as oxidoreductase, transferase, hydrolase, lyase, isomerase, and ligase. The oxidoreductase family of enzymes catalyze the oxidation–reduction reaction, while the transferase catalyzes the transfer of one molecular group to another. The group hydrolases perform the breakage of bonds such as C–O and C–C by hydrolysis reactions. The lyase group of enzymes also perform the breakage of bonds in molecules by using catalytic actions other than hydrolysis and oxidation. Isomerases cause structural changes and molecular rearrangements, while ligases join molecules. The enzymes are immobilized by noncovalent and covalent interactions. Electrostatic adsorption of enzyme has also been reported with good stability. Enzymatic bio-recognition element can be used with optical-, electrochemical-, acoustic-, and calorimetric-type transducers.

Antibodies are a class of bio-affinity-type recognition elements, also called as immunoglobins, used to design biosensors. Antibodies are generally made up of amino acids which are arranged in a particular sequence to form a complex structure. Each antibody is developed to detect a particular antigen and therefore antigen–antibody-based biosensors are highly selective. The antibodies can be monoclonal or polyclonal depending on whether they are made from same or different immune cells. Monoclonal antibodies are expensive and difficult to maintain. The antigen and antibody interact by a lock-and-key model and the interaction force is generally weak in nature such as hydrogen bonding, electrostatic interaction, etc. Hence, the possibility of the reversal of the interaction is also high. Thus, reversible/recoverable biosensors can be designed with antigen–antibody bio-recognition element. To fabricate the sensor, antibodies are immobilized over transducer element surface using methods of amino conjugation, aldehyde, or carboxyl groups after functionalizing the surface. Advantages of antibody–antigen-based sensors include fast response and stability and they also hold disadvantages such as dependency on pH, temperature, environment, and have high cost even for single analyte detection with possibilities of nonregeneration in some cases. The common transducer elements used for such sensors are optical, piezoelectric, electrochemical, and light-addressable potentiometric. For labeling purpose, fluorophores and substrates which are chemically active or complexes are usually used in conjugation with antibody–antigen biosensors.

Microbes are microorganisms that can be used as bio-recognition elements in sensors. The microorganisms are able to metabolize a wide range of chemical substances and complexes and the measurement in the sensor is based on their metabolic activities. Microbial sensors are low cost, stable, and able to carry out complex reactions as compared to the enzymatic and antigen–antibody-based sensors (D'Souza 2001). Their disadvantages are large response time, low sensitivity, and large regeneration time (Karube et al. 1987). The immobilizations of microbes are not performed directly on transducers since they undergo much vigorous reactions. Hence, the microbial recognition element is usually kept little away from the analyzing part and the immobilization is carried out on the cellulose nitrate membrane, nylon net, etc. Due to this, microbes concentration is generally used more than required and is suspended in a solution. Microbes release the enzymes using which they metabolize the analyte. Some microbes alter their respiratory activities while operation and are called as respiratory biosensors. Microbes generally produce a group of enzymes and hence for particular applications, certain microbes are genetically produced that release a single enzyme. Moreover, the microbial cells used can be viable or nonviable (D'Souza 2001). Nonviable cells make use of the enzymatic reactions, while the viable cells make use of metabolic and respiratory functions of the microbe. Viable microbes such as bacteria can be genetically improved to make recombinant organisms. Such organisms can be either immobilized directly to membrane or can be gel entrapped. Supramolecular cell structures such as mitochondria have limited applications in biosensing due to stability issues (Korotkaya 2014).

Nucleic acids are another bio-affinity-based recognition elements. These are larger biomolecules which include RNA, peptides, and DNA. These biomolecules/biopolymers are made from nucleotides, which are the basic monomers. In biosensors utilizing DNA as bio-recognition element, a chain of polymerization reactions is carried out for DNA amplification. These reactions are required to increase the selectivity, applicability, and stability in real-time sensors (Wang et al. 2004). Nucleic acid bio-recognition elements made in a synthetic way are called aptamers. Aptamers can be designed for the binding of various analytes such as proteins, small to large biomolecules, etc. The advantages of aptamer include small size, proper immobilization with stability and reusability. Sensors designed using aptamers are termed as aptasensors. Such sensors are used to find the genetic disorders, toxic microorganism, bacteria, and viruses.

Artificial receptors or synthetic molecular recognition elements or biomimics are also used as bio-recognition layer in sensors. These may include phage display proteins and polypeptides, materials of metal oxides, synthetic proteins, and molecular imprinted polymers. These are low cost, stable, durable, and can be easily fabricated. Immobilization can be performed by modifying the structure. Molecular imprinted polymers are also artificial receptors created on polymers to specifically detect the desired analyte. By using a specific monomer, cross-linker and solvent, polymerization procedure is carried out in the presence of template molecule (analyte). After polymerization, the template molecules are removed using a suitable chemical remover. The procedure leaves the polymer with imprinted sites of the analyte so that only the particular analyte fits into the sites. The method offers high specificity and selectivity of analytes and high sensitivity if implemented over optical fibers with coperforming techniques such as SPR and LMR. The method has applications in sensing of analytes in environment as well as in human body to give sensor probe with high-regeneration capabilities.

As previously discussed, for the realization of biosensors, the recognition element/receptor is immobilized over the transducer, which recognizes the interaction between the analyte and the receptor. The immobilization of the receptors is performed partially or completely over the transducer surface. Nowadays, immobilization of various bio-recognition elements such as enzymes, antibodies, nucleic acids, and peptides is possible. There are several types of immobilization methods to fix the bio-recognition element over the transducer surface. These methods can be categorized into three main groups based on the interaction of the receptor immobilized with the transducer surface. Figure 1.5 shows the schematic of immobilization of the receptors using different methods. These are discussed below.

Adsorption: This is the simplest method of immobilization of the receptors over the transducer surface which covers a broad area of applications with the ability of maximum loading of receptors (nearly 1 g per gram of matrix). In this method, receptor is mixed with a suitable adsorbent under certain specific ambient conditions of ionic strength and pH for a sufficient time interval under incubation. The receptor is then mounted/loaded over the transducer element using an adsorbent which creates a matrix-like structure. The driving force causing this adsorbance is usually the combination of hydrophobic effects and the formation of several salt links per receptor molecule. The selection of adsorbent for the receptor depends upon the iso-electric point difference between the bio-recognition element and the adsorbent matrix and it should prevent the leakage of bio-recognition element during use. The physical links between the enzyme molecules and the support matrix are often very strong; the attachment depends on the type of the substrate used.

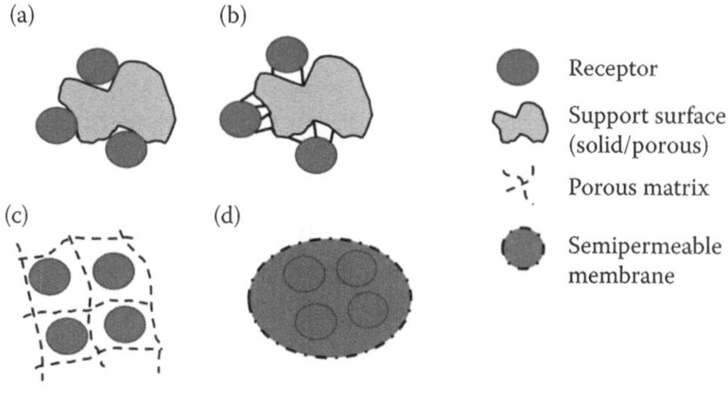

FIGURE 1.5
Schematic of the principal methods of immobilization of receptors in sensing applications. (a) Adsorption, (b) covalent bonding, (c) entrapment, and (d) membrane confinement.

Covalent bonding: As named, in this method, the interaction between the bio-recognition element and the support is of covalent type. Since the covalent bonds are strong by nature, the bonding between the receptor and the support is good enough to prevent the receptor leakage or detachment from the supporting transducer surface. Covalent bonding can be performed by two main routes: one, the receptor is directly attached with the supporting surface through the functional group contained by the receptor, and second, the receptor is attached through the modification in the supporting surface using a specific functional group such that the receptor can bind covalently. Several methods have been reported for immobilization of receptors through covalent binding which depend upon the type of receptor to be immobilized such as enzyme, antibody, and other, and also on the nature of supporting matrix. Some of these are diazotization (SUPPORT–N=N−RECEPTOR), amide bond formation (SUPPORT–CO–NH−RECEPTOR), alkylation and arylation (SUPPORT–CH_2–NH–RECEPTOR and SUPPORT–CH_2–S−RECEPTOR), Schiff's base formation (SUPPORT–CH=N−RECEPTOR), amidation reaction (SUPPORT–CNH–NH−RECEPTOR), and thiol-disulfide interchange (SUPPORT–S–S−RECEPTOR). Although covalent bonding has many beneficial factors such as ease of immobilization, high stability, and no leakage of receptors, the method suffers from loss in receptor activity and degradation in its performance when used repeatedly.

Entrapment: Entrapment of the receptors within the polymeric matrices or in lattice structures is one of the most convenient methods used for the immobilization. The entrapment process is purely a physical caging or involves covalent binding. The medium for the entrapment should have optimized porosity/flexibility so that the target receptor molecule can be stable without any change in its properties. The method is generally used in moderate temperature and pH conditions. The matrix prevents the receptor from degradation. The method provides high-loading capacity but is costly and has diffusion limitations. The difficulty in entrapping large molecule receptors in the catalytic sites limits the application of entrapped receptors with high-molecular-weight substrates.

Membrane confinement: Membrane confinement of the recognition element/receptor can be achieved by variety of methods which depend on the utility and design of the semipermeable membrane. The membrane must confine the receptor while allowing the free passage for the analyte. This is achieved by placing the receptor inside the semipermeable membrane while the analyte moves inside. The analyte interacts with the receptor through the bidirectional diffusion (Mitro et al. 2007). As all the biomolecules are present inside the semipermeable medium, protection of biomolecules from denaturing is ensured at harsh environmental conditions. The method is suitable for the immobilization of enzymes, living cells, microorganisms, etc. (Singh et al. 2013). The method holds a disadvantage of the limitation in controlling the porous size of the membrane which is quite complicated in the case of small biomolecules. Table 1.1 shows a comparison of the immobilization techniques.

TABLE 1.1
General Comparison of Different Immobilization Techniques

Characteristics	Adsorption	Covalent Binding	Entrapment	Membrane Confinement
Preparation	Simple	Difficult	Difficult	Simple
Cost	Low	High	Moderate	High
Binding force	Variable	Strong	Weak	Strong
Enzyme leakage	Yes	No	Yes	No
Applicability	Wide	Selective	Wide	Very wide
Running problems	High	Low	High	High
Matrix effects	Yes	Yes	Yes	No
Large diffusional barriers	No	No	Yes	Yes
Microbial protection	No	No	Yes	Yes

1.6 Biosensor Regeneration

The biosensors designed for each application need to be of low cost, robust, and portable with real-time implementation. Biosensors with multianalyte detection possibility are also getting importance these days, but it makes the sensor costly. If the sensor's cost can be reduced, then it will be affordable and extendable to local markets. In some cases, where the sensor needs to be hygienic to avoid passing of infections and contaminations, sensor is so designed that it can be disposed off. The disposable sensor needs to be cost-effective, which can be ensured by designing the sensing layer using conductive polymeric structure and using disposable transducers. The process of making sensor regenerative or recoverable has advantages as well as disadvantages. Some sensor design requires high-cost elements, high-performance transducer, and high-cost integrated chip as substrate (Ligler 2009). These components cannot be replaced easily and in some cases replacement creates error as well. In such cases, the sensor needs to be regenerative.

The type and period of interaction of the analyte with the bio-recognition element play important roles in deciding whether the sensor is regenerative or not. In biosensor, it is generally difficult to make an active regeneration. If the interaction between biological materials/cells and surface of the sensor is passive, which does not change the morphology or metabolism of the cell, then the surface allows easy recovery resulting in passive regeneration. If the interaction is strong resulting in metabolic changes and morphological alterations of the cell, then the regeneration is active, which makes the regeneration process complicated and difficult to achieve. The type of interaction between the cells/materials and surface can be hydrophilic, hydrophobic, electrostatic, adsorption/absorption, or combination of these. In order to achieve regeneration in a biosensor, it is necessary to overcome the interactive process and forces between the analyte and the bio-receptor element. For this, suitable regeneration buffer can be used.

The regeneration is performed by means of chemical, electrochemical, enthalpic, thermal, and entropic interactions (Goode et al. 2014). Chemical way of regeneration is the most commonly used. In this chemical method, the suspension atmosphere in which the interaction occurs is changed. This is possible by using buffers of suitable pH. The pH change alters the relative charge between the bio-recognition element and the analyte which causes the structural transformation or affects the suspension environment thereby disturbing the interaction between analyte and bio-recognition element, leading to the removal of analyte from the bio-recognition element. This way, the regeneration of the sensor is achieved. The pH of such regeneration buffer varies between 2 and 11. In the case of electrochemical sensor, usually change in electrolyte is used for regeneration purpose. However, the method is limited by the choice of pH and, therefore, it should be ensured that the pH change is not interfering the sensor response and the property of the bio-recognition element. The surfactants such as sodium dodecyl sulfate (SDS) and tween-20 can also be used for this purpose. Another chemical that has been in use in biosensors for easy regeneration with limited damage to bio-recognition element is glycine. Glycine is, generally, used for optical transducer systems. It is a cost-effective regeneration mediator that can be used in acidic environment as well as in aqueous atmosphere because it is dissolvable. Other chemicals that are widely in use include ethylenediaminetetraacetic acid (EDTA), urea, dimethyl sulfoxide, and potassium thiocyanate. Thermal regeneration method is used for sensors utilizing nucleic acid such as DNA, RNA as bio-recognition layer. In the case of electrochemical method, the regeneration can also be performed by applying a negative voltage to the sensor to regenerate the bio-recognition element attached to the transducer. Due to the probability of damaging bio-recognition layer in the electrochemical method of regeneration, it has limited applications. The enthalpic interactions between the analyte and bio-recognition element are altered by the change in the charge interactions between them. This can be brought about by changing the pH, ionic strength, and the suspension environment. The entropy of the system varies depending on the interaction between the analyte and the bio-recognition element. It has been seen that the attachment of analyte on the bio-recognition element decreases the system entropy except for the analytes that are hydrophobic in nature. Thus, the regeneration is also possible by varying the entropy of the system which controls the interaction between analyte and bio-recognition element. There are no defined criteria for the evaluation of regeneration of a biosensor. In general, for a sensor to be regenerative and reproducible, the sensor should provide repeatable results with very less variation/fluctuation.

1.7 Overview of the Book

In this chapter, we have provided a basic idea about the field of sensors. Starting from the definition of the sensor, it provides an overview of the necessities of the sensor in biological, chemical, and environmental areas. It also provides historical developments in the field of sensor and its elements. Various categories of sensors designed using concepts of optics, electrical, chemical, biochemical, thermodynamic, and acoustic aspects have been discussed which brought up the entire reign of sensor community that has got applications ranging from agriculture to space shuttle nowadays. The possibilities and shortcomings of each and every technique provide reader a better choice of the sensor type intended for a particular application.

This book aspires the readers to understand the fundamentals of the sensor and its requisites in day-to-day life. Among the different categories of the sensors discussed, systems based on optical method follow a number of advantages and numerous optical techniques for the realization of the method as well. The compatibility of bio-recognition element is also reported to be favorable with the optical transducers and its detection mechanism. Among the optical substrates that are in use for implementing sensors, the optical fiber is the prime choice due to its peerless properties and compatibility with the systems including facilities of remote sensing, online monitoring, and real-time implementations. Hence, the present book presents the readers an inclusive outlook of the optical sensors with resonance-based techniques that are mainly used for the realization of sensors in biomedical diagnostic and environmental monitoring which are gaining interest and importance these days. In precise, Chapter 2 presents the resonance techniques used for the sensors starting with the basics of different resonance approaches. The well-established resonance technique of SPR has been discussed in detail, along with the theory and realization. In addition, SPR produced in different geometries of the sensor probe design has been discussed. The latest resonance technique named as LMR has been scrutinized with its development in the field of sensing. Chapter 3 deals with the optical-fiber-based sensors, their design, performance parameters, and sensor properties. Tailoring the advantages of optical method of sensing with the modern chemical platforms and materials has resulted in the betterment of sensor's performance and has established a wide application pool. Among these, nanotechnology and nanomaterials are exceptional platforms whose structures modify sensor applications and characteristics, which have been discussed in Chapter 4. Among the materials that hold a stage in the sensor probe fabrication, semiconductor metal oxides are a category with multiple ideal properties that help in gas, chemical, and biosensing. In Chapter 5, the roles of metal oxides in sensing, synthesis, its supporting sensing properties, and the process of sensing have been presented. The application of metal oxides for environmental monitoring has been discussed in detail where the mechanism of LMR has laid its stone in sensing. Chapter 6 gives detailed discussion on the indirect method of sensing using molecular imprinting technique. Classifications in molecular imprinting technique and the idea of using the technique along with SPR and LMR for probe realization as a contemporary approach have been discussed in detail. These are the most recent developments in optical-fiber-based sensor system and have been explained for the first time in a textbook. The concluding chapter provides a summary of the entire discussion of the book with the future scope of work, techniques, and combinations that can be done in the wide field of optical sensors incorporating the optical fiber and resonance techniques.

References

Cammann, K. 1977. Bio-sensors based on ion-selective electrodes. *Fresenius J. Anal. Chem.* 287: 1–9.

Cass, A. E. G., G. Davis, G. D. Francis, H. A. O. Hill, W. J. Aston, I. J. Higgins, E. V. Plotkin, L. D. L. Scott, and A. P. F. Turner. 1984. Ferrocene-mediated enzyme electrode for amperometric determination of glucose. *Anal. Chem.* 56 (4): 667–671.

Clark, L. C. and C. Lyons. 1962. Electrode systems for continuous monitoring in cardiovascular surgery. *Ann. N. Y. Acad. Sci.* 102: 29–45.

Cram, D. J. 1988. The design of molecular hosts, guests, and their complexes (Nobel Lecture). *Angew. Chem. Int. Ed. Engl.* 27: 1009–1020.

D'Souza, S. F. 2001. Microbial biosensors. *Biosens. Bioelectron.* 16: 337–353.

Diamond, D. 1998. *Principals of Chemical and Biological Sensors.* New Jersey, NY: Wiley Interscience Publication.

Durst, R. A., A. J. Baumner, R. W. Murray, R. P. Buck, and C. P. Andrieux. 1997. Chemically modified electrodes: Recommended terminology and definitions. *Pure Appl. Chem.* 69: 1317–1323.

Frant, M. S. 1994. History of the early commercialization of ion-selective electrodes. *Analyst* 119: 2293–2301.

Gauglitz, G. 2005. Direct optical sensors: Principles and selected applications. *Anal. Bioanal. Chem.* 381: 141–155.

Goode, J. A., J. V. H. Rushworth, and P. A. Millner. 2014. Biosensor regeneration: A review of common techniques and outcomes. *Langmuir* 31: 6267–6276.

Grattan, K. T. V. and T. Sun. 2000. Fiber optic sensor technology: An overview. *Sens. Actuators A* 82: 40–61.

Guilbault, G. G. and J. G. Montalvo Jr. 1969. A urea-specific enzyme electrode. *J. Am. Chem. Soc.* 91: 2164–2165.

Gupta, B. D. 2006. *Fiber Optic Sensors: Principles and Applications.* India: NIPA.

Hafeman, D. G., J. W. Parce, and H. M. McConnell. 1988. Light-addressable potentiometric sensor for biochemical systems. *Science* 240: 1182–1185.

Han, X. X., B. Zhao, and Y. Ozaki. 2009. Surface-enhanced Raman scattering for protein detection. *Anal. Bioanal. Chem.* 394: 1719–1727.

Hoffmann, L., M. S. Muller, S. Kramer, M. Giebel, G. Schwotzer, and T. Wieduwilt. 2007. Applications of fiber optic temperature measurement. *Proc. Estonian. Acad. Sci. Eng.* 13: 363–378.

Hulanicki, A., S. Glab, and F. Ingman. 1991. Chemical sensors: Definitions and classification. *Pure Appl. Chem.* 63 (9): 1247–1250.

Jorgenson, R. C. and S. S. Yee. 1993. A fiber-optic chemical sensor based on surface plasmon resonance. *Sens. Actuators B* 12: 213–220.

Karube, I., Y. Wang, E. Tamiya, and M. Kawarai. 1987. Microbial electrode sensor for vitamin B_{12}. *Anal. Chim. Acta* 199: 93–97.

Katz, S. A. and G. A. Rechnitz. 1963. Direct potentiometric determination of urea after urease hydrolysis. *Fresenius J. Anal. Chem.* 196: 248–251.

Keller, U. S. E. 2008. *Chemical Sensors and Biosensors for Medical and Biological Applications.* New Jersey, NY: John Wiley.

Kolthoff, I. M. and H. L. Sanders. 1937. Electric potentials at crystal surfaces, and at silver halide surfaces in particular. *J. Am. Chem. Soc.* 59: 416–420.

Korotkaya, E. V. 2014. Biosensors: Design, classification, and applications in the food industry. *Foods. Raw. Mat.* 2: 161–171.

Kulys, J. J. and G. J. S. Švirmickas. 1980. Reagentless lactate sensor based on cytochrome b_2. *Anal. Chim. Acta* 117: 115–120.

Lehn, J. M. 1988. Supramolecular chemistry-scope and perspectives molecules, supermolecules, and molecular devices (Nobel Lecture). *Angew. Chem. Int. Ed. Engl.* 27: 89–112.

Ligler, F. S. 2009. Perspective on optical biosensors and integrated sensor systems. *Anal. Chem.* 81: 519–526.

Lubbers, D. W. and N. Opitz. 1975. The pCO_2 -/pO_2-optode: A new probe for measurement of pCO_2 or pO in fluids and gases. *Z. Naturforsch. C* 30: 532–533.

Maskow, T., K. Wolf, W. Kunze, S. Enders, and H. Harms. 2012. Rapid analysis of bacterial contamination of tap water using isothermal calorimetry. *Thermochim. Acta* 543: 273–280.

Mello, L. D. and L. T. Kubota. 2002. Review of the use of biosensors as analytical tools in the food and drink industries. *Food Chem.* 77: 237–256.

Mitro, N., P. A. Mak, L. Vargas et al. 2007. The nuclear receptor LXR is a glucose sensor. *Nature* 445: 219–223.

Monosik, R., M. Stredansky, and E. Sturdik. 2012. Biosensors-classification, characterization and new trends. *Acta Chim. Slov.* 5: 109–120.

Moore, C. and B. C. Pressman. 1964. Mechanism of action of valinomycin on mitochondria. *Biochem. Biophys. Res. Comm.* 15: 562–567.

Patel, P. D. 2002. (Bio)Sensors for measurement of analytes implicated in food safety: A review. *Trends Anal. Chem.* 21: 96–115.

Pedersen, C. J. 1967. Cyclic polyethers and their complexes with metal salts. *J. Am. Chem. Soc.* 89: 7017–7036.

Rodriguez-Mozaz, S., M. Marco, M. J. L. De-Alda, and D. Barcelo. 2004. Biosensors for environmental monitoring of endocrine disruptors: A review article. *Anal. Bioanal. Chem.* 378: 588–598.

Reyes, J. I., De-Corcuera, and R. P. Cavalieri. 2003. Biosensors. *Encyclopedia of Agricultural, Food, and Biological Engineering,* eds. D. R. Heldman and C. I. Moraru, 119–123. Boca Raton, FL: CRC Press.

Santos, J. L. and F. Farahi. 2014. Overview of Optical Sensing. *Handbook of Optical Sensors,* eds. J. L. Santos and F. Farahi, 3–11. Boca Raton, FL: CRC Press.

Singh, S., S. K. Mishra, and B. D. Gupta. 2013. SPR based fiber optic biosensor for phenolic compounds using immobilization of tyrosinase in polyacrylamide gel. *Sens. Actuators B* 186: 388–395.

Stefanac, Z. and W. Simon. 1966. Highly selective cation electrode systems based on in-vitro behavior of macrotetrolides in membranes. *Chimica* 20: 436–440.

Thevenot, D. R., K. Toth, R. A. Durst, and G. S. Wilson. 2001. Electrochemical biosensors: Recommended definitions and classification. *Biosens. Bioelectron.* 16: 121–131.

Thompson, R. 2004. *Thunder Underground: Northumberland Mine Disasters*. Derbyshire, England: Landmark Publishing.

Tothill, I. E. 2009. Biosensors for cancer markers diagnosis. *Sem. Cell Dev. Biol.* 20: 55–62.

Updike, S. J. and G. P. Hicks. 1967. The enzyme electrode. *Nature* 214: 986–988.

Wang, R., S. Tombelli, M. Minunni, M. M. Spiriti, and M. Mascini. 2004. Immobilization of DNA probes for the development of SPR based sensing. *Biosens. Bioelectron.* 20: 967–974.

Wang, X. D. and O. S. Wolfbeis. 2013. Fiber-optic chemical sensors and biosensors (2008–2012). *Anal. Chem.* 85: 487–508.

Wang, X., C. Drew, S.-H. Lee, K. J. Senecal, J. Kumar, and L. A. Samuelson. 2002. Electrospun nanofibrous membranes for highly sensitive optical sensors. *Nano Lett.* 2: 1273–1275.

Weinstein, R. L., S. L. Schwartz, R. L. Brazg, J. R. Bugler, T. A. Peyser, and G. V. McGarraugh. 2007. Accuracy of the 5-day freestyle navigator continuous glucose monitoring system-comparison with frequent laboratory reference measurements. *Diabetes Care* 30: 1125–1130.

Wolfbeis, O. S., ed. 1991. *Fiber Optic Chemical Sensors and Biosensors, 1st Volume*. Boca Raton, FL: CRC Press.

2

Basics of Resonance

It has been known that vibrations result in generation of waves in a system. For instance, light waves result by the vibrations of particles that have positive and negative charges, while the sound waves are due to the vibrations occurring mechanically in a system. The natural frequency at which any system or particles tend to vibrate is called its resonant frequency. When sound wave or light wave impinges on a system, the resonant frequency of the light tries to counterpart with that of the system causing an energy transfer from light to the system, termed as resonance. The energy transfer causes the electrons in the system to vibrate at a higher amplitude than before. The system which resonates with light completely absorbs it, while the system that has its electrons not resonating with the impinging light completely transmits the light. A number of sensors have been designed and fabricated using the phenomenon of resonance. In this chapter, we shall discuss various types of resonances and the sensors designed using these resonances.

2.1 Resonance-Based Sensors

Various types of resonance techniques have been used to realize physical, chemical, and biosensors. The transducer part of the sensor makes use of the resonance mechanism to impart energy from the input source to the sensing medium. Among the various kinds of transducer systems employed in a sensor unit and discussed in Chapter 1, the advantages of optical transducers are mainly addressed. A major category among the optical transducers is based on the refractive index (RI) and the technique that has gained major attention in refractive-index-based optical transducers is the surface plasmon resonance (SPR). The phenomenon has been tremendously used in the sensing applications. In an SPR-based sensor, bulk layer of metal is used as the transducer. If, instead of bulk layer, metallic nanoparticles are used, then the phenomenon is called as localized surface plasmon resonance (LSPR). In the next section, we shall discuss the evolution of SPR technique and its application in sensing.

2.2 SPR for Sensing

The phenomenon of SPR was reported by Wood in 1902 when he observed an uneven distribution of light in a reflected spectrum from metal supported diffraction grating (Wood 1902). The idea was not very clear at that time, although he postulated the reason as the interaction between the incident light and the grating–metal configuration. Later, in 1957, Ritchie came up with the term "surface plasmons" (SPs) while reporting the existence of SPs on a thin metallic film surface (Ritchie 1957). In 1960, the excitation of SPs on metal surface using accelerated electrons was outlined (Powell and Swan 1960). Although a primal experimental consideration on SPR was reported in 1959 (Turbadar 1959), a detailed analysis of the condition for SPR generation was communicated by Otto in 1968. He showed the SP excitation in a prism-based configuration with a fall in the reflected power of the light undergoing total internal reflection from the prism base. The configuration consisted of a sandwiched dielectric layer between the prism base and the metal layer (Otto 1968). The necessity of maintaining a miniscule consistent gap between prism base and the metal layer restrained the configuration from its further use. A modified configuration to overcome this limitation was introduced in 1968 itself by Kretschmann and Raether by removing the miniscule gap between the prism base and the metal layer, with a simple and easy configuration named as Kretschmann configuration (Kretschmann and Reather 1968). A detailed discussion on these configurations is given in the further sections.

The technique of SPR started exploring its application since 1970 showing its potential for the characterizations of thin films and their monitoring processes (Pockrand et al. 1978). The gas sensing and biosensing applications of SPR were initially reported by Liedberg and coworkers (Liedberg et al. 1983) and the first experimental results of SPR-based chemical sensor were reported by Jorgenson and Yee (1993). Due to enormous advantages such as fast response, label-free detection, high sensitivity and selectivity, SPR has gained much attraction in recent years in the field of sensing (Homola et al. 1999, Usha et al. 2016). Below, we shall first discuss in detail the theories of SPR and LSPR and then the realization of sensors based on these techniques. To differentiate between SPR and LSPR, SPR occurs on the excitation of SPs at a metal–dielectric interface, while in the LSPR the localization of electric field around the metallic nanostructures is observed. The new optical technique of lossy mode resonance (LMR) and its development for sensing purposes will be discussed later in this chapter.

2.2.1 Theory of SPR

Before understanding the physics of SPR, it is necessary to understand plasmons and SPs. Plasmons are the quantum of wave-like excitations in the plasmonic systems where the electrons/ions interact with each other through Coulomb interactions. It can be found in metals, highly doped semiconductors, ionic gases, etc. In metals, the Coulomb interaction between the free electrons and the positive-charge background takes place which is discussed as the Jellium model. When an electric field is applied on metal at a point, the local free electron density is reduced due to Coulomb repulsion resulting in an increased positive-charge background at the same point. An increase in positive-charge background attracts the free electrons again because of Coulomb attraction. Due to the attraction of free electrons to the positive-charge background, the local negative charge increases and a restoring force in terms of Coulomb repulsion produces, because of which the free electrons start repulsing. This process of repulsion–attraction continues and results in a longitudinal oscillation of free electrons. These oscillations are termed as plasma oscillations and the quanta of these oscillations are called as plasmons. The plasma oscillation can be visualized as the wave propagation in water. If we drop a small particle in calm water, a longitudinal wave is produced which propagates away from its origin which is similar to the propagation of the plasma waves (Gupta et al. 2015). The frequency of plasmons, ω_p, is expressed as

$$\omega_p = \sqrt{\frac{Ne^2}{\varepsilon_0 m}} \tag{2.1}$$

where N, e, ε_0, and m correspond to the electron density, charge of an electron, electric permittivity of vacuum, and the mass of a free electron, respectively.

Plasmons are generally classified in three categories: volume plasmons, SPs, and localized surface plasmons (LSPs). Volume plasmons are plasmons which are generated in the bulk quantity of metals due to the external electric field and the separation of charges within the metal. SPs are produced at the metal–dielectric interface, while the LSPs are the confinement of SPs in the nanosized particles.

The Maxwell equations in the case of no external charge and current density are written as

$$\nabla \cdot \vec{D} = 0 \tag{2.2}$$

$$\nabla \cdot \vec{B} = 0 \tag{2.3}$$

$$\nabla \times \vec{E} = -\frac{\partial \vec{B}}{\partial t} \tag{2.4}$$

$$\nabla \times \vec{H} = \frac{\partial \vec{D}}{\partial t} \tag{2.5}$$

where $\vec{E}$, $\vec{D}$, $\vec{H}$, and $\vec{B}$ are the electric field, displacement vector, magnetic field, and magnetic induction, respectively. Using $\vec{D} = \varepsilon_0 \varepsilon \vec{E}$ and $\vec{B} = \mu_0 \vec{H}$, these equations give following wave equations

$$\nabla^2 \vec{E} - \frac{\varepsilon}{c^2} \frac{\partial^2 \vec{E}}{\partial t^2} = 0 \tag{2.6}$$

$$\nabla^2 \vec{H} - \frac{\varepsilon}{c^2} \frac{\partial^2 \vec{H}}{\partial t^2} = 0 \tag{2.7}$$

where ε_0 and μ_0 are the electric permittivity and magnetic permeability of vacuum, respectively; ε is the dielectric constant or relative permittivity; and c is the velocity of light in vacuum. For the understanding of surface plasmons, we consider a semi-infinite metal–dielectric interface as shown in Figure 2.1. Further, we consider wave propagation along the x-direction (metal–dielectric interface) at $z = 0$. The time-dependent wave propagating along the x-axis with propagation constant β and angular frequency ω can be written as

$$\vec{E} = \vec{E}_0 e^{i(\beta x - \omega t)} \tag{2.8}$$

$$\vec{H} = \vec{H}_0 e^{i(\beta x - \omega t)} \tag{2.9}$$

Substituting these equations in equations 2.4 and 2.5, one obtains the following equations for the system shown in Figure 2.1:

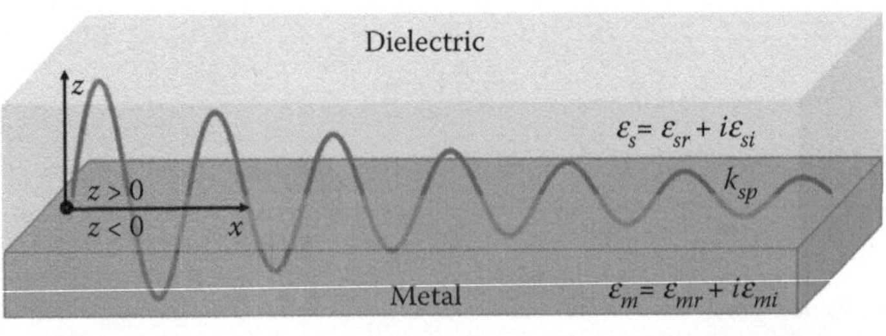

FIGURE 2.1
Schematic of the propagation of surface plasmon wave at the metal–dielectric interface.

$$
\begin{aligned}
& i\beta E_y = i\omega\mu_0 H_z \\
& \frac{\partial E_y}{\partial z} = -i\omega\mu_0 H_x \\
& \frac{\partial H_x}{\partial z} - i\beta H_z = -i\omega\varepsilon_0\varepsilon E_y \\
& i\beta H_y = -i\omega\varepsilon_0\varepsilon E_z \\
& \frac{\partial H_y}{\partial z} = i\omega\varepsilon_0\varepsilon E_x \\
& \frac{\partial E_x}{\partial z} - i\beta E_z = -i\omega\mu_0 H_y
\end{aligned}
\tag{2.10}
$$

Out of these six equations, first three equations are solely for the components (E_y, H_x, H_z) and the other three are for the components (H_y, E_x, E_z). In other words, first three equations are independent of the other three. First three equations define transverse electric (TE) mode, while the other three define the transverse magnetic (TM) mode. In TE mode, there does not exist electric field in the direction of propagation and the magnetic field is perpendicular to *E*-field. Similarly, in the case of TM modes, there is no magnetic field in the direction of propagation while the electric field is always perpendicular to the *H*-field.

For TE and TM modes, the wave equations can be obtained from the first three and the second three equations, respectively. These are

For TE polarization:

$$\frac{d^2E_y}{dz^2} + \left(k_0^2\varepsilon - \beta^2\right)E_y = 0 \tag{2.11}$$

For TM polarization:

$$\frac{d^2H_y}{dz^2} + \left(k_0^2\varepsilon - \beta^2\right)H_y = 0 \tag{2.12}$$

where $k_0 = \omega/c$.

Let us first consider TM polarization. The solution of wave equation 2.12 with evanescently decaying field along the z-axis can be written as under.

For $z > 0$

$$H_y(z) = A_1 e^{i\beta x} e^{-k_1 z} \tag{2.13}$$

For $z < 0$

$$H_y(z) = A_2 e^{i\beta x} e^{k_2 z} \tag{2.14}$$

Use of equation 2.10 and the following boundary conditions

$$H_y(z > 0) = H_y(z < 0)$$

$$\varepsilon_1 E_z(z > 0) = \varepsilon_2 E_z(z < 0)$$

give

$$A_1 = A_2 \tag{2.15}$$

Applying the following boundary condition

$$E_x(z > 0) = E_x(z < 0)$$

and using equation 2.15, one obtains

$$\frac{k_1}{k_2} = -\frac{\varepsilon_1}{\varepsilon_2} \tag{2.16}$$

Since k_1 and k_2 are positive, equation 2.16 implies that the dielectric constants of the two media should be of opposite signs if a wave is to propagate along the interface. Thus, one of the media should be a metal, while the other is a dielectric.

Substitution of equation 2.13 in equation 2.12 gives

$$k_1^2 = \beta^2 - k_0^2\varepsilon_1 \tag{2.17}$$

While the substitution of equation 2.14 in equation 2.12 gives

$$k_2^2 = \beta^2 - k_0^2\varepsilon_2 \tag{2.18}$$

Using equation 2.16, the propagation constant of SP wave can be obtained from equations 2.17 and 2.18 for metal–dielectric interface as

$$\beta = k_{sp} = k_0\left[\frac{\varepsilon_{mr}\varepsilon_s}{\varepsilon_{mr} + \varepsilon_s}\right]^{1/2} \tag{2.19}$$

where we have substituted $\varepsilon_1 = \varepsilon_{mr}$, which is the real part of the complex dielectric constant ($\varepsilon_m = \varepsilon_{mr} + i\varepsilon_{mi}$) of the metal and $\varepsilon_2 = \varepsilon_s$ is the dielectric constant of the dielectric medium in contact of metal.

Now we consider TE polarization. The solution of wave equation 2.10 with evanescently decaying field along the *z*-axis can be written as under.

For $z > 0$

$$E_y(z) = A_1 e^{i\beta x} e^{-k_1 z} \tag{2.20}$$

For $z < 0$

$$E_y(z) = A_2 e^{i\beta x} e^{k_2 z} \tag{2.21}$$

For TE modes, boundary conditions are

$$H_x(z > 0) = H_x(z < 0)$$

$$E_y(z > 0) = E_y(z < 0)$$

The combination of these conditions gives

$$A_1(k_1 + k_2) = 0 \tag{2.22}$$

For the propagation of the light waves through the interface, Re (k_1) and Re (k_2) should be positive which is possible only for $A_1 = 0$; hence, $A_1 = A_2 = 0$. Thus, it can be concluded that the SP mode does not exist for TE polarization. A pictorial representation of propagation of the SP wave is shown in Figure 2.1.

A plot between k_{sp} and ω as the dispersion curve of the SPs for metal–dielectric interface is plotted in Figure 2.2. In the same figure, solid line

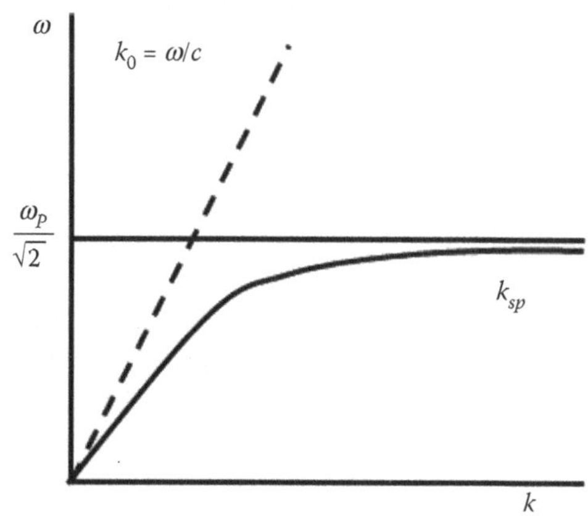

FIGURE 2.2
Dispersion curves of surface plasmon wave and direct light.

shows the propagation vector of SPs, while dotted line shows the propagation vector for direct light.

The propagation length of the SP wave at wavelength λ_0 is given as

$$\delta_{spw} = \frac{\lambda_0}{2\pi} \frac{\varepsilon_{mr}^2}{\varepsilon_{mi}} \left(\frac{\varepsilon_{mr} + \varepsilon_s}{\varepsilon_{mr}\varepsilon_s} \right)^{3/2} \tag{2.23}$$

Further, the penetration depths of SP wave in metal are expressed as

$$\delta_m = \frac{\lambda_0}{2\pi} \left[\frac{\varepsilon_{mr} + \varepsilon_s}{\varepsilon_{mr}^2} \right]^{1/2} \tag{2.24}$$

and the penetration depths of SP wave for the dielectric medium are expressed as

$$\delta_s = \frac{\lambda_0}{2\pi} \left[\frac{\varepsilon_{mr} + \varepsilon_s}{\varepsilon_s^2} \right]^{1/2} \tag{2.25}$$

For instance, in the case of silver–air interface, use of plasma and collision frequencies of silver as $\omega_p = 1.2 \times 10^{16}$ rad/s and $\omega_c = 9.1 \times 10^{13}$ rad/s gives the values of propagation length and penetration depths in silver and air for 400 nm wavelength as 11.29 μm, 24.58 nm, and 134.8 nm, respectively (Srivastava and Gupta 2013). The values show that penetration depth of SP wave is small for metal than the dielectric. This is because of the faster field decay in metals. Penetration depth in dielectric corresponds to the length over which the SPs are sensitive for the change in RI of the dielectric, while in the case of metal, penetration depth is the minimum thickness of the metal for the realization of SPs. Since the SPs are TM polarized, these can be excited via TM or p-polarized light. In other words, the SPs propagating along the metal–dielectric interface can be excited by the incident light having same propagation vector and the polarization as that of SPs. If we look at the dispersion curves in Figure 2.2 for direct light and SP wave, it can be easily understood that the propagation vector of direct light and surface plasmons can never be equal, which concludes that the direct light cannot excite SPs. For an excitation of SPs, the propagation vector of the direct light should be increased. This can be achieved if the light is incident at the metal–dielectric interface after passing through a medium having high RI so that the propagation vector of the light is increased. For this, initially two distinct prism-based configurations were proposed by Otto and Kretschmann–Reather in 1968, named as Otto configuration and Kretschmann configuration, respectively. These two configurations are schematically shown in Figure 2.3.

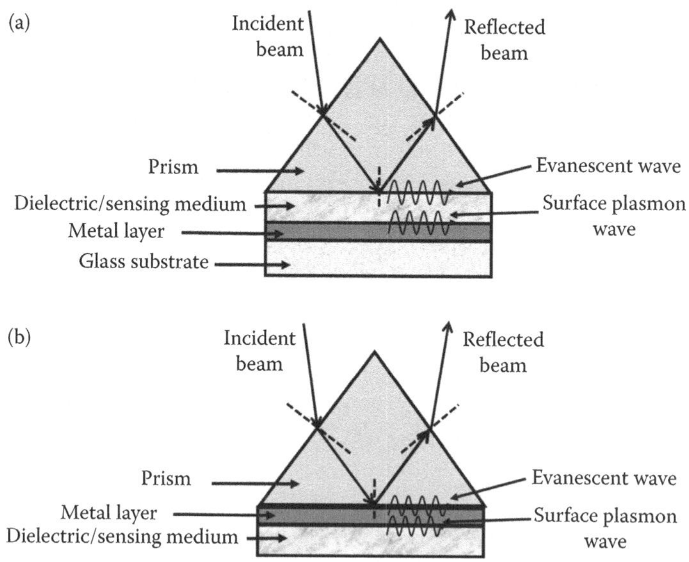

FIGURE 2.3
(a) Otto and (b) Kretschmann configuration for realizing the SPR.

Figure 2.3a represents the Otto configuration, while Figure 2.3b shows the Kretschmann configuration. In both the cases, TM-polarized light is launched from a monochromatic source through one of the faces of the prism with an incident angle greater than the critical angle at the prism base. The reflected light is then collected by the power meter at the other face of the prism as shown in the figure. For the angle of incidence greater than the critical angle, the total internal reflection of the light occurs and an evanescent wave is generated, which propagates along the interface with an exponentially decaying field in the rarer medium. The propagation vector (k_{ev}) of the evanescent wave is written as

$$k_{ev} = k_0 \sqrt{\varepsilon_p} \sin\theta \tag{2.26}$$

where ε_p and θ are the dielectric constant of the prism and the angle of incidence at the prism base, respectively. In Otto configuration, a metal layer is coated over the glass substrate and an infinitesimal (few nanometer-sized) gap is kept between the metal layer and the prism base. The gap is filled by the dielectric material having the dielectric constant (ε_s) such that $\varepsilon_s < \varepsilon_p$. When light is launched in the way as shown in Figure 2.3a, the SPs at the metal–dielectric interface get excited by the evanescent wave produced at the prism–dielectric interface. When the propagation vectors of both the waves become equal, the incident light energy is completely transferred to the excitation of SPs. This condition of resonance is satisfied at a particular angle of incidence called as resonance angle (θ_{res}), which leads to a sharp dip in the reflected light intensity at θ_{res} measured by varying the angle of incidence. The phenomenon is called as the SPR. The Otto configuration is generally used for the characterization of the metal film properties. However, this configuration is not of much use in applications as it is difficult to maintain the nanosized gap between the metal and the prism base. In the case of Kretschmann configuration, a few nanometer-thick film of metal is coated over the prism base followed by a layer of a dielectric medium with dielectric constant ε_s such that $\varepsilon_s < \varepsilon_p$; the schematic of which is shown in Figure 2.3b. For this configuration, the evanescent wave produced at the prism base–metal interface due to the total internal reflection propagates along the interface and excites the SPs at the metal–dielectric interface. When the propagation vectors of these two waves match, the resonance occurs. Mathematically, the resonance condition can be written as

$$k_{sp} = k_0 \left[\frac{\varepsilon_m \varepsilon_s}{\varepsilon_m + \varepsilon_s} \right]^{1/2} = k_0 \sqrt{\varepsilon_p} \sin\theta_{res} = k_{ev} \tag{2.27}$$

The matching of the propagation vectors can be graphically understood from Figure 2.4 showing the crossing of the dispersion curves of SP wave and evanescent wave.

From equation 2.27, it can be noted that the resonance angle depends on the dielectric constants of prism, metal, and dielectric/sensing medium. For a given prism and metal layer, any change in the dielectric constant or RI of the dielectric/sensing medium changes the resonance angle. As the RI of the sensing medium increases, the resonance angle increases. This property of SPR is mainly used for the sensing applications. In other way, if the angle of incidence is kept constant and the polychromatic light source is used in the configuration, then in the SPR spectrum recorded after reflection from prism–metal interface, a dip (or the resonance) occurs at a particular

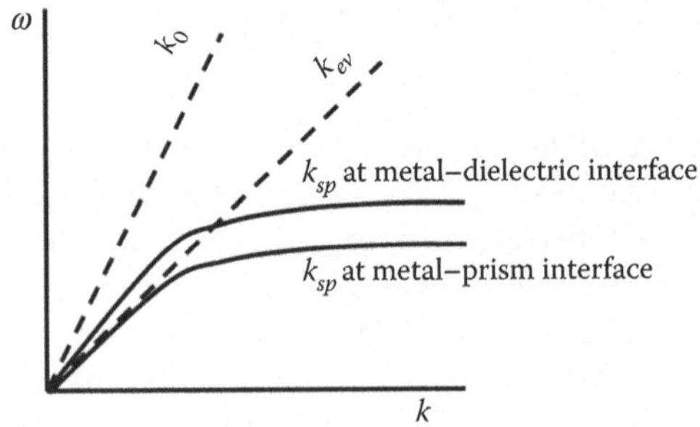

FIGURE 2.4
Dispersion curves of surface plasmon waves at the metal–dielectric and prism–metal interfaces and the evanescent wave in the Kretschmann configuration.

wavelength called as the resonance wavelength λ_{res}. An increase in the RI of the dielectric/sensing medium increases the resonance wavelength. The sensor realization using these properties is further discussed in Section 2.2.3. It may also be noticed that prism–metal interface also supports the generation of SPs but these cannot be excited by the direct incident light or evanescent wave. This is because of the higher propagation vector of the SP wave at the prism–metal interface than that of the evanescent wave as shown in Figure 2.4. Other types of configurations used for the realization of SPR are based on waveguide and optical fiber. The configuration in these is similar to the Kretschmann configuration except that the prism is replaced by the waveguide layer or the core of the fiber. The metal layer is coated over the small portion of the waveguide/unclad fiber core and a dielectric/sensing medium is kept in contact with the other surface of the metal layer. To achieve excitation, the dielectric constant of the sensing medium should be less than that of the waveguide layer/core of the fiber. In the case of waveguide/fiber core, light is guided by means of total internal reflection and hence the evanescent wave is produced at the interface which excites the SPs (generated at the metal–dielectric interface). The schematics of the waveguide- and optical-fiber-based SPR configurations are shown in Figure 2.5a and b, respectively.

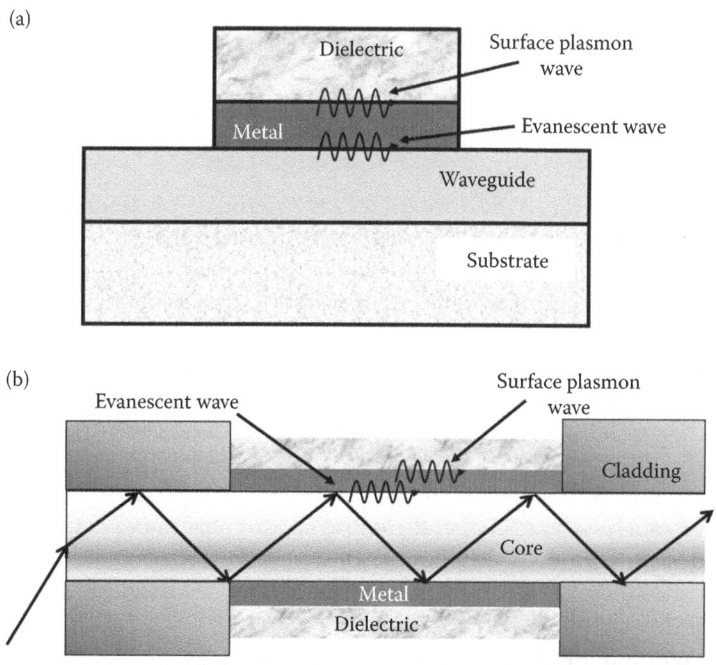

FIGURE 2.5
Realization of SPR using (a) waveguide and (b) optical-fiber-based configurations.

Carrying out the angular interrogation scheme in a fiber/waveguide-based configuration is quite cumbersome due to the broad range of guided waves supported by the fiber and emerging from the output end of different angles. Hence, in these cases, usually polychromatic light source is used for the excitation of SPs and a dip in the output transmitted spectrum is observed at the resonance wavelength. It may also be noted that, in the case of optical fiber, instead of TM-polarized light, unpolarized light is used because, due to the cylindrical geometry of fiber and bending of the fiber core, maintaining a single polarization in the fiber core is difficult. The unpolarized light has both TE and TM modes and as discussed earlier, SPR is achieved only by TM modes and TE modes do not make any effects on SPs. Thus, the output light received is a combination of unaffected TE modes and the TM light which is used for SPR achievement. Therefore, the light observed at the output end has the minimum value of normalized transmitted power as 0.5. The SPR curves observed for fiber-based configuration show much broader curves compared to the prism-based configuration due to the increased number of reflections within the sensing region of the fiber core. The shift in resonance wavelength, as mentioned earlier, due to the change in the RI of the dielectric/sensing medium is used for the realization of the SPR-based fiber-optic sensors. It has been discussed in details in Section 2.2.3.

The confinement of SPs in metallic nanoparticle causes their localization and is termed as LSPs. In the case of nanoparticles, the resonance occurs at a particular frequency and is called LSPR. In the next section, we shall discuss briefly the LSPs.

2.2.2 Theory of LSPR

The LSPs are produced due to the confinement of plasmons in metallic nanoparticle with size comparable to or smaller than the wavelength of incident light. In the presence of the external electric field, metallic nanoparticle behaves as an oscillating dipole. At resonance, the frequencies of incident light and collective oscillations of conduction electrons in metallic dipole match. This results in an increase in the extinction of light by metallic nanoparticle and the phenomenon is called LSPR (Mayer and Hafner 2011). The LSP has two important effects: Electric fields near the particle's surface are greatly enhanced and the particle's optical absorption has a maximum at the resonance wavelength. The resonance wavelength depends on various parameters such as size, shape, dielectric constant of nanoparticles and the surrounding medium. The dependency on the RI of surrounding medium provides the role of LSPR in sensing, which shall be discussed in the next section of this chapter.

To understand the theory of LSPR, we must cognize the interaction of light with a small particle that may result in scattering or absorption of light by the particle. The combination of scattering and absorption is called as extinction. Initially, we shall start by the quasi-static (electrostatic) approximation, where the scattering (C_{sca}), absorption (C_{abs}), and extinction (C_{ext}) cross sections, for a small particle, are given as (Bohren and Huffman 1998)

$$\begin{aligned} C_{sca} &= \frac{k^4}{6\pi\varepsilon_0^2}\,|\,\alpha^2\,|, \\ C_{abs} &= \frac{k}{\varepsilon_0}\operatorname{Im}(\alpha), \\ C_{ext} &= C_{sca} + C_{abs} \end{aligned} \tag{2.28}$$

where $k = 2\pi\sqrt{\varepsilon_s}/\lambda$, ε_s is the dielectric constant of the surrounding medium and α is the polarizability of the particle. To calculate these coefficients, we first derive the expression for the polarizability of a single molecule of spherical shape with radius R. An external electric field ($\vec{E}_{ext}$) is applied to the particle and due to this electric field a local electric field ($\vec{E}_{loc}$) is induced which is the sum of the three electric fields as given below:

$$\vec{E}_{loc} = \vec{E}_{ext} + \vec{E}_{int} + \vec{E}_{L} \tag{2.29}$$

Here, $\vec{E}_{int}$ is the electric field inside the nanoparticle and $\vec{E}_L$ is the electric field produced due to the polarization of particle. Figure 2.6 shows pictorial representation of field applied over the metal nanoparticle.

Since net electric field inside a metal is zero, we can write

$$\vec{E}_{int} = 0 \tag{2.30}$$

The external electric field, $\vec{E}_{ext}$, causes a parallel polarization vector $\vec{P}$ on the surface of the metal nanoparticle having surface charge density σ given by

$$\sigma = \vec{P}\cdot\vec{n} = -P\cos\theta \tag{2.31}$$

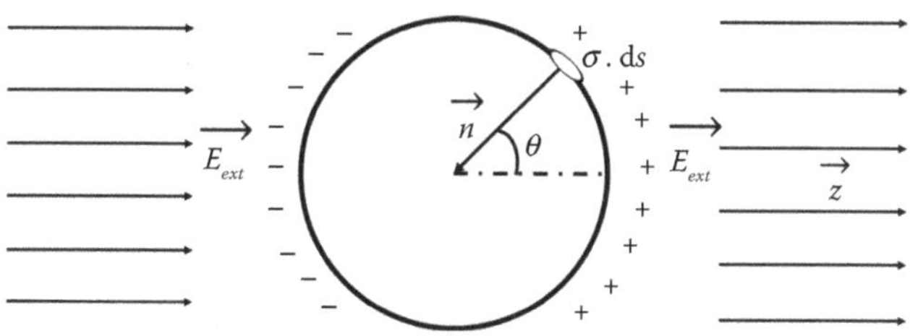

FIGURE 2.6
Metal nanoparticle in external electric field ($\vec{E}_{ext}$).

where $\vec{n}$ is the normal vector of surface element toward center and θ is the angle between the electric field direction and surface element for small area dS. If we concentrate on the small surface area dS, the total charge for the surface element would be σdS and the corresponding electric field according to the Coulomb's law is

$$E = \frac{1}{4\pi\varepsilon_0}\frac{\sigma dS}{R^2} \tag{2.32}$$

Its value along the direction of $\vec{E}_{ext}$ (if $\vec{z}$ is the unit vector) is

$$\vec{E} = \frac{1}{4\pi\varepsilon_0}\frac{\sigma dS}{R^2}\cos\theta\vec{z} \tag{2.33}$$

Integrating equation 2.33 for the whole surface of nanoparticle, we obtain

$$\overrightarrow{E_L} = -\oint \vec{E} = \frac{-1}{4\pi\varepsilon_0}\oint \frac{\sigma dS}{R^2}\cos\theta\vec{z} = \frac{1}{4\pi\varepsilon_0}\oint \frac{PdS}{R^2}\cos^2\theta\vec{z} \tag{2.34}$$

By using the relation $dS = R^2 \sin\theta d\theta d\phi$ (using spherical coordinates),

$$\begin{gathered}\overrightarrow{E_L} = \frac{1}{4\pi\varepsilon_0}\int_0^{2\pi} d\phi \int_0^{\pi} \frac{P}{R^2}\cos^2\theta R^2 \sin\theta d\theta\vec{z} = \frac{2\pi}{4\pi\varepsilon_0}\int_0^{\pi} P\cos^2\theta \sin\theta d\theta\vec{z} \\ \overrightarrow{E_L} = \frac{P}{2\varepsilon_0}\left[-\frac{\cos^3\theta}{3}\right]_0^{\pi}\vec{z} = \frac{P}{3\varepsilon_0}\vec{z}\end{gathered} \tag{2.35}$$

(negative prefix is because of the direction of the field component σdS, which is toward negative $\vec{z}$-direction).

Substituting the values of $\overrightarrow{E_L}$ and $\vec{E}_{int}$ from equations 2.35 and 2.30 in equation 2.29, we get

$$\overrightarrow{E_{loc}} = \overrightarrow{E_{ext}} + \frac{\vec{P}}{3\varepsilon_0} \tag{2.36}$$

By definition, polarization ($\vec{P}$) and polarizability (α) are related as

$$\vec{P} = \frac{\alpha\vec{E}_{loc}}{V} \tag{2.37}$$

where V is the volume of the nanoparticle.

From equations 2.36 and 2.37, we can write by rearranging the terms

$$\vec{P} = \frac{3\alpha \vec{E}_{ext}}{3\varepsilon_0 V - \alpha} \tag{2.38}$$

By definition, susceptibility (χ) is related to polarization vector as $\vec{P} = \varepsilon_0 \chi \vec{E}_{ext}$. Substituting this in equation 2.38, we can write susceptibility as

$$\chi = \frac{3\alpha}{3\varepsilon_0 V - \alpha} \tag{2.39}$$

Using $1 + \chi = \varepsilon_m/\varepsilon_s$ in equation 2.39, where ε_m and ε_s are the dielectric constants of metal nanoparticle and surrounding medium, respectively, the polarizability can be written as

$$\alpha = 3V\varepsilon_0 \left[\frac{\varepsilon_m - \varepsilon_s}{\varepsilon_m + 2\varepsilon_s} \right]$$

or

$$\alpha = 4\pi R^3 \varepsilon_0 \left[\frac{\varepsilon_m - \varepsilon_s}{\varepsilon_m + 2\varepsilon_s} \right] \tag{2.40}$$

According to equation 2.28, to achieve resonance or maximum extinction, α should be maximum; and according to equation 2.40, α is maximum if the value of $|\varepsilon_m + 2\varepsilon_s|$ is minimum. Thus, the resonance condition for LSPR can be achieved if $\mathrm{Re}[\varepsilon_m] = -2\varepsilon_s$. This is also known as Fröhlich condition and the corresponding modes supporting this condition are called dipole SP modes (Maier 2007). The frequency corresponding to these modes is $\omega = \omega_p/\sqrt{3}$. Referring to equations 2.28 and 2.40, the scattering, absorbance, and extinction cross sections can be written as

$$C_{sca} = \frac{8\pi}{3} k^4 R^6 \left| \frac{\varepsilon_m - \varepsilon_s}{\varepsilon_m + 2\varepsilon_s} \right|^2 \tag{2.41}$$

$$C_{abs} = 4\pi k R^3 \,\mathrm{Im} \left| \frac{\varepsilon_m - \varepsilon_s}{\varepsilon_m + 2\varepsilon_s} \right|$$

or

$$C_{abs} = \frac{24\pi^2 \varepsilon_s^{3/2} R^3}{\lambda} \frac{\mathrm{Im}(\varepsilon_m)}{[\mathrm{Re}(\varepsilon_m) + 2\varepsilon_s]^2 + [\mathrm{Im}(\varepsilon_m)]^2} \tag{2.42}$$

According to Equation 2.28, extinction cross section (C_{ext}) is the sum of the scattering cross section and the absorption cross section. While the values of C_{sca} and C_{abs} depend upon k^4R^6 and kR^3, respectively, and for the nanoparticle of radius (R) around 10 nm $k^4R^6 \ll kR^3$ and therefore, $C_{sca} \ll C_{abs}$. Thus, after neglecting the C_{sca}, the expression of extinction cross section (C_{ext}) can be written as

$$C_{ext} = \frac{24\pi^2\varepsilon_s^{3/2}R^3}{\lambda} \frac{\mathrm{Im}(\varepsilon_m)}{[\mathrm{Re}(\varepsilon_m)+2\varepsilon_s]^2+[\mathrm{Im}(\varepsilon_m)]^2} \tag{2.43}$$

The transmitted light intensity after its interaction with nanoparticle can be written as

$$I = I_0(\omega) - \frac{I_0(\omega)}{S} C_{ext} \tag{2.44}$$

where $I_0(\omega)$ is the intensity of the input light and S is the surface area of the nanoparticle. We have now understood the theory of LSPR by quasi-static approximation (QSA) for a single spherical nanoparticle. The expression of α varies for different shapes of the nanostructures and their dielectric constants. Further, there are several theories reported for the understanding of LSPR beyond QSA approximation such as Mie theory which gives the expression for polarizability as (Mie 1908)

$$\alpha = \varepsilon_0 V \left[\frac{1 - \frac{1}{10}(\varepsilon_m + \varepsilon_s)x^2 + O(x^4)}{\left(\frac{1}{3} + \frac{\varepsilon_s}{\varepsilon_m - \varepsilon_s}\right) - \frac{1}{30}(\varepsilon_m + 10\varepsilon_s) - i\frac{4\pi^2\varepsilon_m^{3/2}}{3}\frac{V}{\lambda^3} + O(x^4)} \right] \tag{2.45}$$

where x is the size parameter and its value is equal to $\pi R/\lambda$. O stands for higher-order terms. We can observe from equations 2.41 through 2.43 that the condition of resonance depends on various parameters such as shape, size, geometry, dielectric constants of metal nanoparticle, and the surrounding medium. This makes LSPR a suitable tool for sensing applications. The theoretical analyses of SPR and LSPR show that the resonance condition depends on the dielectric constants of the surrounding medium and the metallic nanostructure or thin film. In the next section, we shall discuss how this dielectric constant dependency of resonance behavior is used for realizing the sensors.

2.2.3 Realization of Sensors

The sensors utilizing SPR/LSPR technique rely on the change in the RI of the sensing medium around the metal thin film/nanostructure. The metals used in SPR/LSPR configuration are the transducing layers. The change in the dielectric constant of the surrounding sensing medium causes the change in various properties of SPR/LSPR curves. Depending on these properties and the oriented application, various interrogation schemes are used for designing the sensors. Common interrogation schemes include angular interrogation, spectral interrogation, intensity interrogation, and phase interrogation. First two schemes are the most commonly used schemes for sensor designing and therefore, here, we shall discuss about these two schemes in detail. Angular interrogation is mostly used for the SPR sensors based on prism configuration. In this scheme, a monochromatic light source is incident on the base of the prism coated with a thin film of metal through one of the faces of the prism. The intensity of the reflected light is collected through the other face as a function of angle of incidence of the light beam. A sharp dip at a particular angle of incidence is obtained, named as resonance angle, for a given RI of the sensing medium kept in contact of the other side of the metal layer. Figure 2.7 shows the SPR curves for three different RIs of the sensing medium. The curves depict that the resonance angle increases with an increase in the RI of the surrounding/sensing medium. A plot of resonance angle as a function of RI of the surrounding/sensing medium is the calibration curve of the sensor. The calibration curve is used to estimate the RI of an unknown sample by measuring the resonance angle.

Spectral interrogation is another widely used interrogation scheme for the realization of SPR-based sensors. In this scheme, the angle of incidence is kept fixed and a polychromatic light source is used. The spectrum of the reflected light from the face of the prism is recorded for a given sensing medium. In the spectrum, a dip is obtained at a particular wavelength and

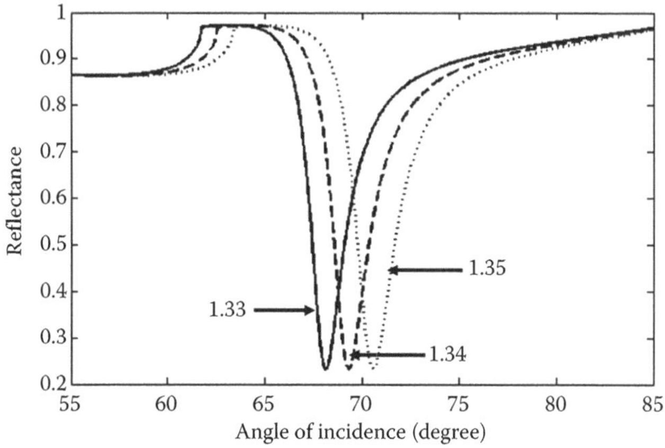

FIGURE 2.7
SPR spectra in the case of prism-based configuration with 633 nm wavelength of light, 40 nm silver film thickness and 1.732 refractive index of the prism for varying RI of the sensing medium.

is called as the resonance wavelength. The resonance wavelengths are determined for samples of different RIs and are plotted as a function of RI. The curve is the calibration curve for this scheme. In this scheme, the resonance wavelength increases with an increase in the RI of the surrounding/sensing medium. The wavelength interrogation method is more advantageous than the angular interrogation method because it does not use any mechanical moving parts to scan light source/detector for a range of angles as required in the case of angular interrogation. Waveguide/fiber-based configurations also use spectral interrogation scheme since these support the propagation of light over a range of angles. In the intensity interrogation method, the source is monochromatic and the angle of incidence is kept constant. The intensity of the reflected light is measured as a function of RI of the surrounding medium at a fixed wavelength. In phase interrogation method, the phase difference between the s-polarized and p-polarized lights is obtained as a function of the RI of the surrounding medium. The phase-interrogation-scheme-based sensor configuration possesses very high sensitivity to power fluctuations. Due to this limitation, its usage is limited in practical sensing applications.

To study the dependency of SPR behavior on metals and surrounding medium RI, a simple fiber-optic configuration shown in Figure 2.5b is considered. An unclad core region on the optical fiber has a coating of a plasmonic metal of suitable thickness. The effects of various parameters such as plasmonic metals (Ag, Au, Al), thickness of metal layer and fiber geometry (core diameter, sensing length, numerical aperture [NA]), and the change in the surrounding medium RI on SPR behavior can be studied using the N-layer model shown in Figure 2.8.

N-layer model is a transfer matrix method, which is used to determine the reflection coefficient of an N-layer structure. A generic jth layer has the thickness d_j, RI n_j, permeability μ_j, and dielectric constant ε_j. The tangential components of the electromagnetic fields at the initial layer boundary, U_1 and V_1, are related to the components of the final layer boundary, U_{N-1} and V_{N-1}, as

$$\begin{pmatrix} U_1 \\ V_1 \end{pmatrix} = M \begin{pmatrix} U_{N-1} \\ V_{N-1} \end{pmatrix} \tag{2.46}$$

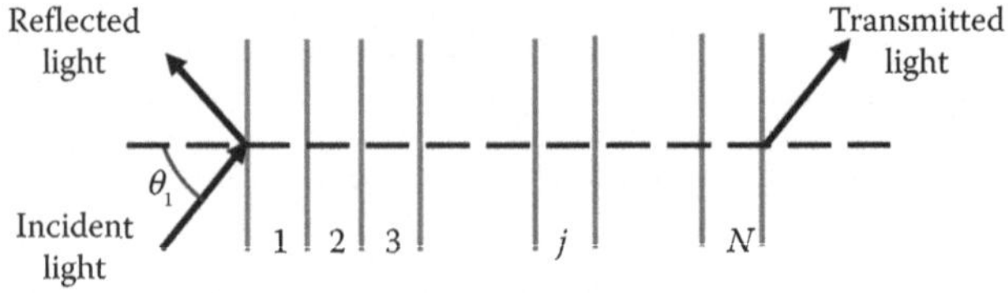

FIGURE 2.8
N-layer model to analyze the transmittance for a fiber-based Kretschmann configuration.

where M is the specific matrix with a combinational structure for the propagating light as

$$M = \prod_{j=2}^{N-1} M_j = \prod_{j=2}^{N-1} \begin{pmatrix} \cos\beta_j & \dfrac{-i\sin\beta_j}{q_j} \\ -i\beta_j \sin\beta_j & \cos\beta_j \end{pmatrix} \tag{2.47}$$

and

$$\beta_j = \frac{2\pi d_j}{\lambda}\sqrt{\varepsilon_j - n_1^2 \sin^2\theta_1} \tag{2.48}$$

while the value of q_j depends on the polarization of light. For TE polarized light, its value is

$$q_j = \sqrt{\varepsilon_j - n_1^2 \sin^2\theta_1} \tag{2.49}$$

and for TM-polarized light

$$q_j = \sqrt{\frac{\mu_j}{\varepsilon_j}}\cos\theta_j = \frac{\sqrt{\varepsilon_j - n_1^2 \sin^2\theta_1}}{\varepsilon_j} \tag{2.50}$$

The transfer matrix, M_j, can be calculated by substituting the values of β_j and q_j in equation 2.47. The total transfer matrix M for a multilayer structure can be obtained by multiplying the transfer matrices for each independent layer. After simplification and rearranging all the terms, one can obtain the amplitude reflection coefficient for an N-layer structure as

$$r = \frac{(M_{11} + M_{12}q_N)q_1 - (M_{21} + M_{22}q_N)}{(M_{21} + M_{22}q_N) + (M_{11} + M_{12}q_N)q_1} \tag{2.51}$$

The reflection intensity coefficient can be obtained from r as

$$R = |r|^2 \tag{2.52}$$

For a polarized light having either TE polarization or TM polarization, R is calculated individually by using the respective expression of q_j as given in equations 2.49 and 2.50 and may be termed as R_{TE} and R_{TM}. In the case of unpolarized light, the value of R is calculated as

$$R = \frac{R_{TE} + R_{TM}}{2} \tag{2.53}$$

The angular power distribution of the rays inside the optical fiber for the collimated beam focused using a microscope objective on the face of the fiber is given by

$$dP \propto \frac{n_{core}^2 \sin\theta \cos\theta d\theta}{(1 - n_{core}^2 \cos^2\theta)^2} \tag{2.54}$$

where n_{core} is the RI of the fiber core and θ is the incident angle of the guided ray with respect to the normal to core–cladding interface. Using equations 2.53 and 2.54, the normalized transmitted power at the output end of the fiber is evaluated as

$$P_{trans} = \frac{\int_{\theta_c}^{\pi/2} R^{N(\theta)} \frac{n_{core}^2 \sin\theta \cos\theta}{(1 - n_{core}^2 \cos^2\theta)^2} d\theta}{\int_{\theta_c}^{\pi/2} \frac{n_{core}^2 \sin\theta \cos\theta}{(1 - n_{core}^2 \cos^2\theta)^2} d\theta} \tag{2.55}$$

where $\theta_c = \sin^{-1}\left(\frac{n_{clad}}{n_{core}}\right)$ is the critical angle and $N(\theta) = \frac{L}{2\rho \tan\theta}$ is the number of reflections in the unclad sensing region with L as the sensing length and ρ as the radius of the fiber core.

Figure 2.9a shows the SPR curves of an optical fiber sensor using different plasmonic metals (Ag, Au, and Al) for the surrounding medium RI of 1.33. The thickness of the metal layer is 40 nm and the dielectric constants of these metals are determined using the Drude model. From the figure, it can be observed that the Al possesses the sharpest SPR curve, while Au exhibits broadest curve. This is because of the highest and lowest values of imaginary part of the dielectric constant of Al and Au, respectively.

Figure 2.9b shows the effect of varying thickness of Ag film on the SPR curves for surrounding RI of 1.33 in the case of prism-based SPR configuration for 55.61° angle of incidence. The change in the metal film thickness changes the depth and slightly the position of the SPR curve. The minimum transmittance is observed for Ag film thickness of 50 nm, which implies the exact matching of propagation vectors around 50 nm.

Figure 2.10a shows the SPR curves for the varying RI of surrounding medium from 1.33 to 1.40 in the case of 40-nm-thick silver layer. It can be seen that an increase in RI causes a redshift of resonance wavelength. The resonance wavelengths extracted from each SPR curve are plotted as a function of surrounding medium RI in Figure 2.10b. This is the calibration curve of the sensor. This can be used to estimate the RI of an unknown sample if its resonance wavelength is experimentally derived. The derivative of the calibration curve is the sensitivity of the sensor and is plotted in Figure 2.10c for different RIs of the surrounding

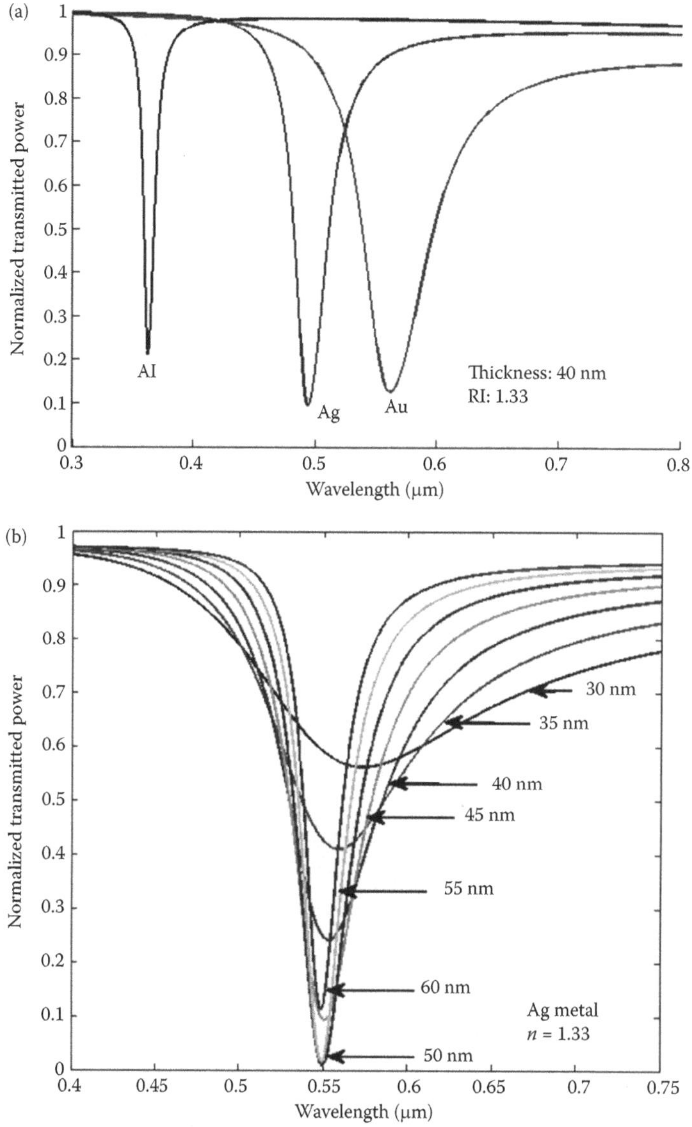

FIGURE 2.9
SPR spectra for (a) different metals coated over the fiber core with core diameter 200 μm and NA 0.20, and (b) varying thickness of silver layer coated over the base of the prism of RI 1.732.

medium. The sensitivity of an SPR-based sensor depends upon the metal used. The metal having high value of real part of the complex dielectric constant possesses high sensitivity. Thus, among Au and Ag, the sensor with Au film possesses high sensitivity because the real part of its dielectric constant is greater than the real part of the dielectric constant of Ag (Gupta and Verma 2009).

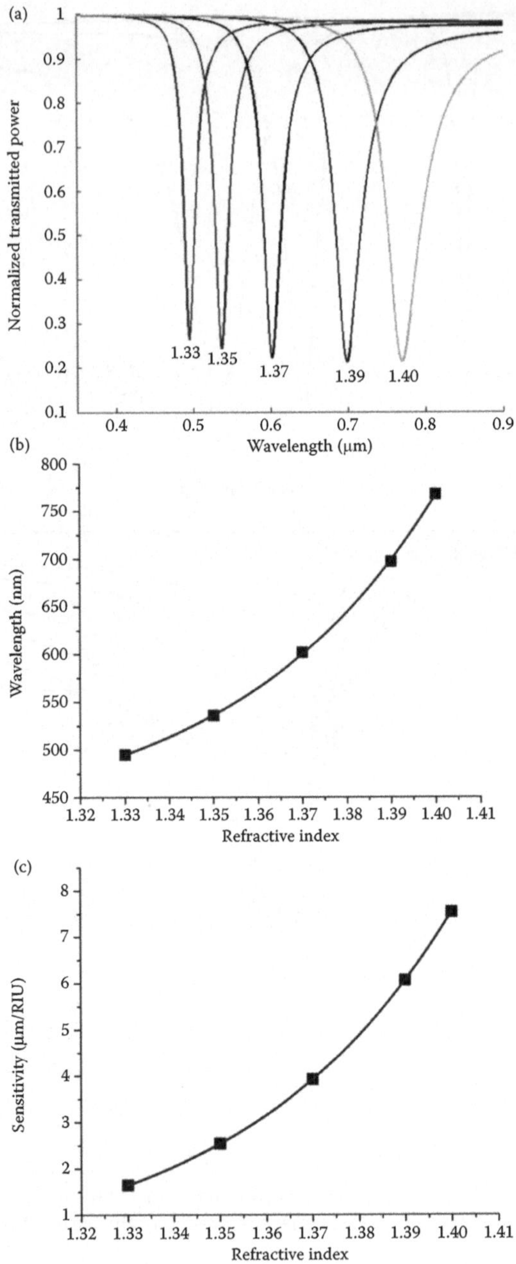

FIGURE 2.10
(a) SPR spectrum, (b) calibration curve, and (c) FOM of a fiber-optic refractive index sensor configuration using Ag (40 nm) as the plasmonic metal for generic fiber-optic refractive index sensor. The plots are for 40 nm thickness of silver layer, 600 μm core diameter, 0.20 NA, and 1 cm sensing length.

The performance of the SPR sensor can be evaluated comprehensively if one takes into account the full width at half maximum (FWHM) of the SPR curve in addition to the sensitivity of the sensor. The performance parameter that takes into account both of these is termed as the figure of merit (FOM). It is the ratio of the sensitivity and the FWHM. The sensor with maximum FOM possesses the best performance.

The effects of core diameter and sensing length of the fiber on the FOM of the sensor are shown in Figure 2.11a and b, respectively. The FOM increases with an increase in the core diameter, but decreases as sensing length increases. This is because the FWHM depends on the number of reflections in the sensing region of the fiber. As the number of reflections increases, the FWHM also increases. Therefore, if the core diameter increases, then the number of reflections decreases and FWHM decreases which increases FOM. Similarly, if the sensing length increases, then the number of reflections increases which decreases the FWHM and hence the FOM. There is a negligible effect of core diameter and sensing length on the sensitivity of the sensor. Figure 2.11c shows the effect of NA of the fiber on the FOM of the sensor. As NA increases, the FOM decreases because an increase in NA of fiber decreases the critical angle causing a broad range of angles of the guided rays in the fiber which results in an increased number of reflections inside the core. Thus, the FWHM of the SPR curve increases and FOM decreases as NA increases.

The sensing performance of an SPR-based sensor can be improved by including a few nanometer thick over layer of high-index material such as silicon (Si) and metal oxides (ZnO, SnO_2). The high-index layer increases the evanescent field intensity at the dielectric and the sensing medium interface which in turns enhances the sensitivity of the sensor (Shalabney and Abdulhalim 2010). This type of SPR configuration is usually termed as nearly guided wave SPR (NGWSPR) and has been discussed in Section 2.3.1.3.

2.3 Types of Resonances

Optical resonance can be achieved by using various kinds of configurations that can be used for realizing sensors. The sensitivity and FOM of such sensors can be improved by rearranging the layers in configuration and by adding the extra layers such as of high-index materials. Below, we discuss various types of resonance techniques and configurations.

2.3.1 Surface Plasmon Resonance

2.3.1.1 Long-Range SPR

The performance of the SPR-based sensor can be improved by increasing the sensitivity and maintaining the certainty in the resonance position. To

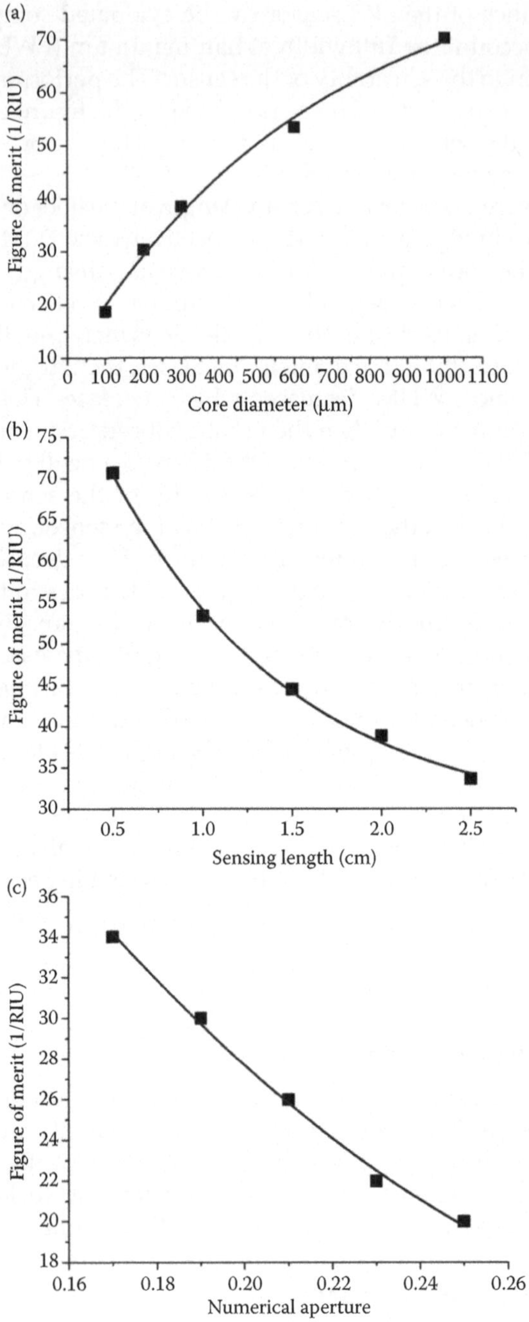

FIGURE 2.11
Effect of (a) core diameter, (b) sensing length, and (c) numerical aperture on FOM of the SPR-based fiber-optic sensor.

reduce the uncertainty in the position of resonance, it is necessary to reduce the width of the SPR curve, which finally improves the sensor resolution. A way to improve the resolution and sensitivity of the sensor is by using long-range surface plasmon resonance (LRSPR) (Nenninger et al. 2001). This can be achieved when a thin metal film is kept between the two dielectric layers having the same dielectric constant.

The SPR discussed in the above section shows the generation of SP modes in the Kretschmann configuration for a thin layer of metal placed between the prism and the sensing/analyte medium (dielectric) having dissimilar RIs. In the same configuration, if a dielectric layer having the same RI as the sensing medium is introduced between the metal film and prism, then for the two interfaces: metal-sensing medium and metal–dielectric, two SP modes can be generated. Figure 2.12a shows the Kretschmann configuration for two SP modes generation on the either side of the metal film, sandwiched between two dielectric layers. The coupling of the two SP modes is possible only when the metal film thickness is thin enough. When these two plasmon modes superimpose in phase, a symmetric mode arises termed as the long-range SP mode. Figure 2.12b shows the long-range SP mode in a dielectric–metal–dielectric configuration.

The attenuation of long-range surface plasmon polaritons (LRSPPs) is at least two times lesser than that of the surface plasmon polaritons in a single interface. This causes the LRSPP to propagate over a longer distance and hence the name long-range SPP (Berini 2009). The lower propagation loss makes LRSPR a better choice for the sensing purposes.

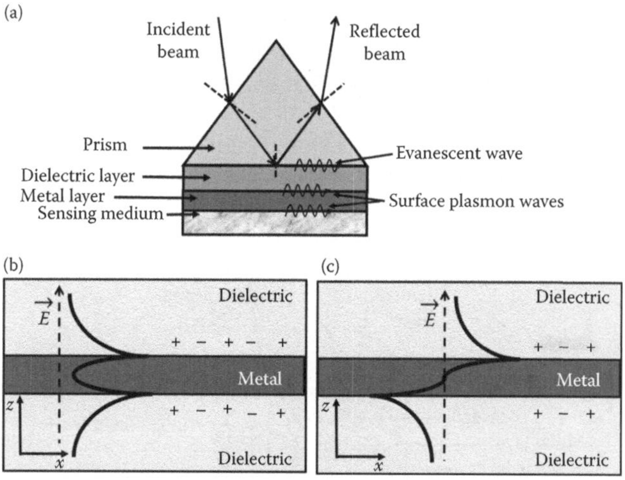

FIGURE 2.12
(a) The Kretschmann configuration with generation of surface plasmons on both sides of metal film sandwiched between two dielectric layers, (b) long-range SPR, and (c) short-range SPR.

2.3.1.2 Short-Range SPR

The same configuration shown in Figure 2.12a can be used for the short-range surface plasmon resonance (SRSPR). If there is an out-of-phase superimposition of the two SPs modes generated on either sides of the metal film, sandwiched between two dielectrics, then an antisymmetric mode arises. This antisymmetric mode is termed as short-range SP mode. Figure 2.12c shows the short-range SP mode in the dielectric–metal–dielectric configuration.

A combination of generation of LRSPR and SRSPR is called dual-mode SPR. In dual mode, the system can be made sensitive separately to RI variations on the surface and bulk of the structure. This is possible because two modes possess different sensitivities to bulk and surface disturbances in the structure. Such dual-mode sensors can efficiently realize a system to characterize between the binding interactions occurring at the surface of sensing layer with the analyte and the background perturbations (happening in bulk), which is an added advantage over single SPR mode sensor system. This is achieved by making use of the difference in the penetration depths of the two modes.

2.3.1.3 Nearly Guided Wave SPR

By introducing a dielectric layer of high RI between the metal layer and sensing medium in the simple SPR configuration, nearly guided wave surface plasmon modes (NGWSPMs) can be generated as shown in Figure 2.13. The thickness of the dielectric layer is chosen in such a way that it supports the guided wave which results in an asymmetric metal-clad waveguide where both TM and TE modes exist (Lahav et al. 2008). Employing a thin dielectric layer with a high value of the real part of the RI on the metal film enhances the amplitude of electric field intensity at the dielectric–sensing medium interface. This enhancement in field results in a high interaction between the SP mode and the sensing medium. Hence, an increment in the sensitivity is possible in NGWSPR sensors compared to the simple SPR sensors.

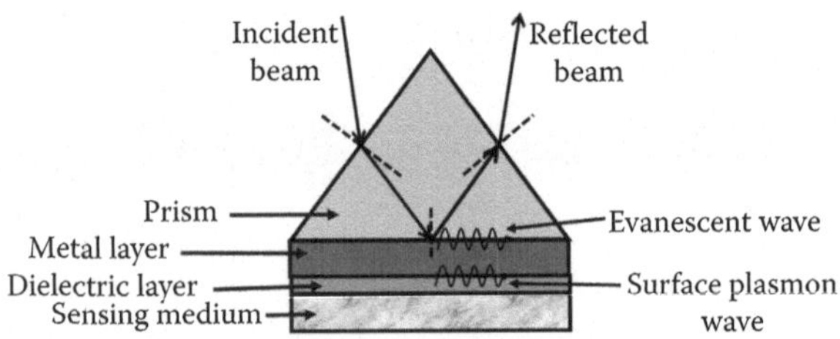

FIGURE 2.13
Schematic of generation of nearly guided wave SPR.

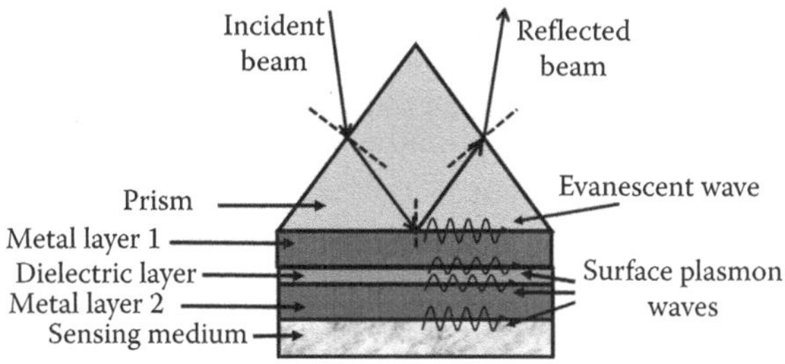

FIGURE 2.14
Schematic of waveguide-coupled-prism-based SPR configuration.

2.3.1.4 *Waveguide-Coupled SPR*

The waveguide-coupled SPR is realized by keeping a dielectric waveguide sandwiched between two metal layers to create a two metal–waveguide interface. This structure is mainly used for tuning the resonance wavelengths from visible to infrared region and for the generation of multiple resonance dips which may be selectively used for suitable applications. Figure 2.14 shows the prism-based configuration for realizing the waveguide-coupled SPR.

The multiple dips correspond to the coupling of different waveguide modes with the SP modes produced at two metal–waveguide interfaces. The values of resonance wavelengths can be controlled by optimizing the thickness of the dielectric waveguide inserted between the metal layers.

2.3.1.5 *Magneto-Optic SPR*

Magneto-optic SPR (MOSPR) is a plasmonic resonance phenomenon realized by combining SPR and magneto-optic Kerr effect (MOKE). A prism-based configuration for achieving MOSPR is shown in Figure 2.15.

A few nanometer-thick layer of magnetic material (e.g., cobalt, Co) is coated between two metal layers as shown in the figure. The dielectric/sensing medium is in the vicinity of the second metal layer. The excitation of SPs occurs at the metal–dielectric interface with the evanescent wave generated

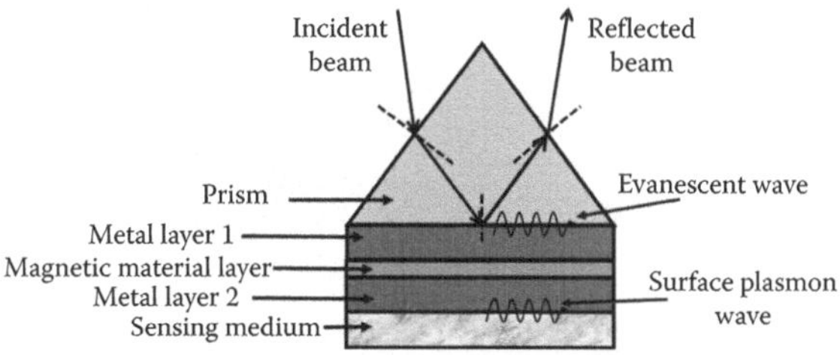

FIGURE 2.15
Magneto-optic SPR configuration realized on prism substrate.

at the prism–metal interface. The presence of magnetic layer results in MOKE effect, which enhances the evanescent field and electromagnetic field of the SP wave. The combined effect results in the MOSPR phenomenon. It has been reported that the sensitivity of MOSPR is about four times greater than SPR, which can further be enhanced by optimizing the properties of the magnetic material layer sandwiched between the metal layers (Kämpf et al. 2012).

2.3.1.6 Fano Resonance SPR

Fano resonance is a resonant-scattering process, where the superimposition of two scattering amplitudes occurs. One is due to the bulk/background process and the other due to the excitation of the resonance phenomenon. The combination of fano resonance with SPR to realize optical sensors for practical applications enhances the sensitivity of the sensor by around 2000 times in comparison to simple SPR configuration. The arrangement is shown in Figure 2.16. It is similar to the Kretschmann configuration. In this arrangement, the base of the prism is coated with a metal layer followed by a coupling layer (of suitable dielectric constant) that can support the generation of SPs at the metal–coupling layer interface. Over the coupling layer, a guiding layer is introduced which can be treated as a planar waveguide and is surrounded by the sensing medium. The light incident on the prism–metal interface generates evanescent wave whose field decays exponentially in to the metal layer to excite SP wave. The generated SP–evanescent field tail can excite planar waveguide mode. This planar waveguide mode is followed by an evanescent field in the coupling layer and sensing layer. The generation of evanescent fields in the coupling layer due to planar waveguide mode and SP mode results in a coupling between them. The degree of coupling and hence the sensor performance can be controlled by optimizing the thickness and the RI of the coupling and guiding layers. Using the Kretschmann configuration, a coupling between the SP mode and planar waveguide mode can be achieved to generate a hybrid mode. The reflectance spectrum at the output shows the appearance of sharp fano resonance in the SPR curve (Zheng et al. 2017).

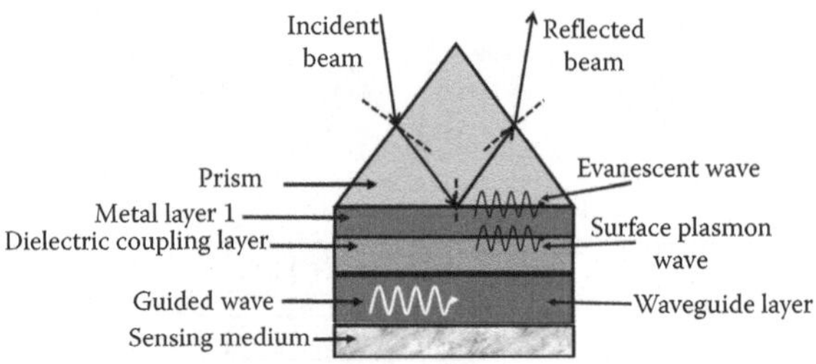

FIGURE 2.16
Prism-based configuration combining fano resonance and surface plasmon resonance.

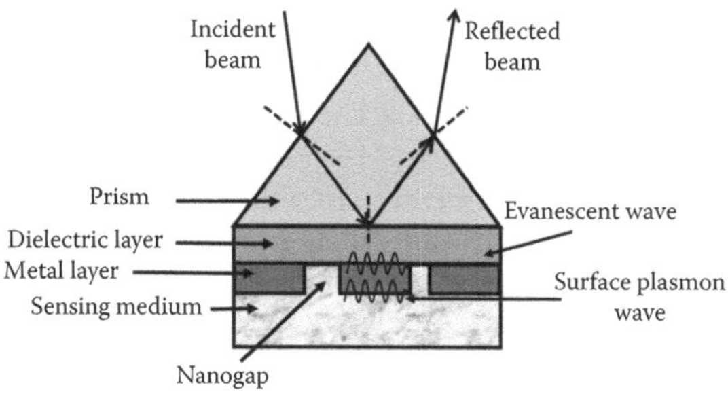

FIGURE 2.17
Gap plasmon realization on prism-based dual-mode SPR configuration.

2.3.1.7 Gap SPR

In a dual-mode sensor, sensitivity enhancement can be achieved by creating the cavities or periodic gaps in the metal layer (Keathley and Hastings 2012). Plasmons that exist in small metallic gaps, which are occupied partially or completely with dielectric materials, are termed as gap plasmons. They exhibit good localization in a direction perpendicular to the gap (Pile et al. 2005). A nanogapped SPR in a dual-mode configuration is shown in Figure 2.17. Standing waves are formed at the pocket-like gaps and the plasmon waves generated are a mixture of propagating and localized waves (Keathley and Hastings 2012). Gap plasmons are assumed to withstand structural disturbances and imperfections. This is because of their high transmission through sharp edges since the bend losses and leakage losses into metals are impossible (Pile et al. 2005).

2.3.2 Lossy Mode Resonance

The phenomenon of resonance using a layer of semiconductor metal oxide or polymers in place of metal layer in the simple Kretschmann configuration is called LMR. The metal oxides/polymers having positive real part of dielectric constant over substrate of lower dielectric constant support lossy mode in them. The LMR-based sensor provides freedom of choice of material to support LMR and the range of operation of the sensor in the electromagnetic spectrum. Multiple LMR dips generation is a unique property of the phenomenon that supports high-sensitive sensing applications. The LMR phenomenon supporting the sensing and sensor realization on optical fiber configuration is discussed in detail in Section 2.4.

2.3.3 Interferometric Resonance

When light beams from two different coherent sources or from a single source having same or complementary phase meet at a point, interference occurs. The interference may be constructive or destructive in nature, depending on the phase difference between the sources. Constructive interference leads to a bright spot, while the destructive leads to a dark spot. This is because the intensities of the two beams are added up (due to the same phase) in constructive interference, while in destructive interference the intensities get subtracted (due to the opposite phase). The condition of interference depends on the optical path difference between the two light beams. If there is any change in optical path of one of the beams, a change in the interference pattern occurs. Hence, if we keep a transparent optical sample in the path of one of the light beams, the optical path difference between the beams changes, resulting in a change in the interference pattern. Thus, by observing the change in the interference pattern, one can determine the RI of the sample. The change in interference pattern provides the information about the change in various properties of the light beam such as wavelength, phase, intensity, and so on, which are correlated to the RI of the sample. This can be used for realizing the sensors using interferometric phenomenon. The interferometric sensors can be designed with high sensitivity and accuracy.

The term "interferometric resonance" instead of "interference" is also used in the literature since the resonance occurring due to the coupling of two waves looks similar to the interference which occurs due to the superposition of two light waves. The fiber interferometers are more advantageous in comparison to conventional ones since the fiber provides resistance to external interference. In fiber-optic interferometers, light can be coupled and propagated through single or multiple fibers. There are four types of fiber-optic interferometric sensor designs such as Fabry–Perot, Michelson, Mach–Zehnder, and Sagnac. In this section, we shall discuss two of these because of their easy design and simple construction.

2.3.3.1 Fabry–Perot Interferometric Resonance

The Fabry–Perot interferometric resonance occurs due to the interference of multiple light beams. In the Fabry–Perot configuration, two partially reflecting mirrors are placed in such way that when light is incident on one mirror it gets reflected from it and moves toward another mirror. The light reflected from this mirror again falls on the first mirror and this back and forth between the two reflecting mirrors goes on. This results in the superposition of multiple transmitted beams or the resonance between the light beam transmitted from a mirror. At resonance, dips in the transmitted/reflected spectrum are observed depending on the type of interrogation used. The positions of these dips change while changing the optical properties of the medium between the partially reflecting mirrors. This is the mechanism behind sensors using the Fabry–Perot interference/resonance (MacPherson 2014). The generic setup of fiber-optic-based Fabry–Perot interferometric sensor is designed by building the reflectors inside or outside the fiber. Thus, the classification can be made as intrinsic Fabry–Perot and extrinsic Fabry–Perot interferometric sensor. Intrinsic type of sensor contains reflecting components within the fiber, while the extrinsic type contains a cavity external to the fiber where the back and forth reflection of light occurs between the partially reflecting mirrors of the cavity (Lee et al. 2012). The schematics of the extrinsic and intrinsic fiber-optic interferometric sensor configurations are shown in Figure 2.18a and b, respectively. The gap between the reflecting components is used as the sensing medium for the realization of sensors. The extrinsic type holds advantages of simplicity and low-cost design but disadvantages of low coupling power, handling, and packaging issues limit its use. Intrinsic-type sensors require expensive fabrication equipment for the cavity formation.

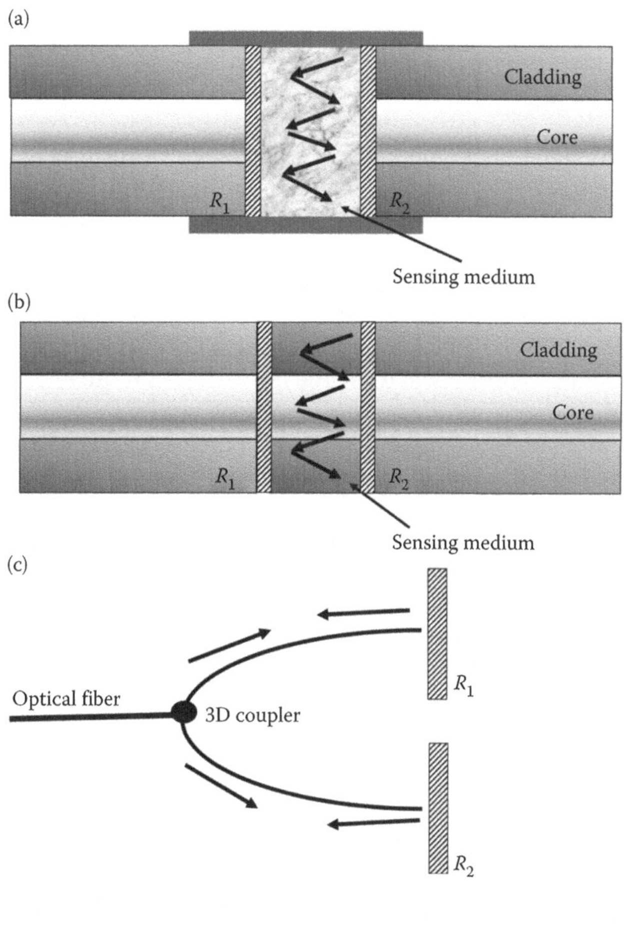

FIGURE 2.18
Schematic diagram of (a) extrinsic, (b) intrinsic Fabry–Perot, and (c) Michelson interferometric resonance configurations.

2.3.3.2 Michelson Interferometric Resonance

Michelson interferometer is the most commonly used interference method. The sensor based on the Michelson interferometer is quite similar to the Fabry–Perot resonance. In this configuration, the monochromatic light propagating in a single fiber is split into two and the two light waves traveling through two different paths get reflected from mirrors and superimpose to generate interference/resonance pattern (MacPherson 2014). In this type of interferometer, sensor is realized by putting an unknown sample in the path of one of the splitted/reflected light beams. The presence of sample changes the path length traveled by the light and hence the interference pattern. By observing the change in the interference pattern, one can find the RI or other properties of the sample. The schematic of a fiber-optic Michelson interferometer is shown in Figure 2.18c (Lee et al. 2012). One arm of the interferometer is used as the reference beam, while in the other arm the sample is kept and hence is the sensing beam.

2.3.4 Acoustic Resonance

Acoustic resonance phenomenon in an acoustic system that amplifies the sound whose frequency matches with one of the natural frequencies of the system is termed as resonant frequency. Usually, there exist several resonant frequencies for an acoustic system. The change in mass, conductivity, elasticity, and the dielectric properties of the resonant system changes the values of resonant frequencies. This property supports the sensing application using acoustic waves. The acoustic waves generated at the surface of the system are termed as surface acoustic waves, while if these are produced in the bulk of the system then they are called as bulk acoustic waves. In the case of optical sensors, coupling of SP waves and surface acoustic waves is used (Friedt and Francis 2016). For example, in biosensing applications using this configuration, the binding of analyte molecules with the active sites present at the recognition medium causes change in the optical and mechanical properties of the sensing unit. The change in the optical properties is detected by SPR by observing the shift in resonance wavelength/angle in the output spectra, while the change in mechanical properties is detected by observing the change in resonant frequency. Thus, by combining both the phenomena, a highly sensitive sensor can be achieved as it evaluates the actual binding of analytes with the recognition sites.

2.4 Developing LMR for Sensing Applications

Mechanism of sensing is a salient factor that decides the role of sensor in any field. Among the optical techniques which are of current interest and included in the book, LMR is an advancing technique that has gained researcher's attraction in a very short duration since its emergence. The main attribute to this is the wide range of the sensor, low cost, and easily available materials that support LMR phenomenon. The materials that support LMR include semiconducting metal oxides (SMO), polymers, organometallics, and any dielectric material that exhibits a lossy nature or satisfies the conditions of LMR. Since metal oxides hold the advantage of abundant material on the earth with ease of availability, research on LMR-based sensors has mainly been focused on SMO-based LMR designs.

In LMR phenomenon, the leakage of electric field propagating in the guided substrate of dielectric in nature to a semiconductor medium occurs at a cutoff thickness. The leaked modes or the modes that are lost from the main guiding substrate region to the semiconductor medium are termed as lossy modes. The maximum leakage occurs at a particular value of the resonance parameter and the phenomenon at this resonance parameter is called as LMR. The LMR configuration on optical fiber substrate shown in Figure 2.19 is a replica of the Kretschmann configuration on fiber-based SPR probe discussed in Figure 2.5b. Over an unclad fiber region, a layer of SMO is coated. The SMO layer is surrounded by a dielectric/sensing medium. Light from an unpolarized light is coupled to the core of the fiber. Due to the leakage of the guided mode at the core–SMO interface, the movement of charge carriers in the SMO is affected. As we know a change in the mobility of the charge carrier changes the conductivity of the SMO and hence its permittivity (Marciniak et al. 1993). A change in permittivity can also be depicted as a change in the material RI. RI of a material is an optical property that is dependent on the wavelength of the incident light. Thus, the wavelength at which the maximum change in the RI occurs is evaluated as the resonance wavelength (λ_{res}). Hence, in the region of SMO-clad fiber, a redistribution of guided modes occurs and their resonance characteristic depends on the RI of the SMO.

To implement the LMR phenomenon for RI sensing, the arrangement having SMO over unclad fiber at the central region is used as the sensor probe. A change in the RI of the surrounding sensing medium of the probe changes the condition of LMR and in result changes the resonance wavelength. A plot between the resonance wavelength and the RI of the surrounding sensing

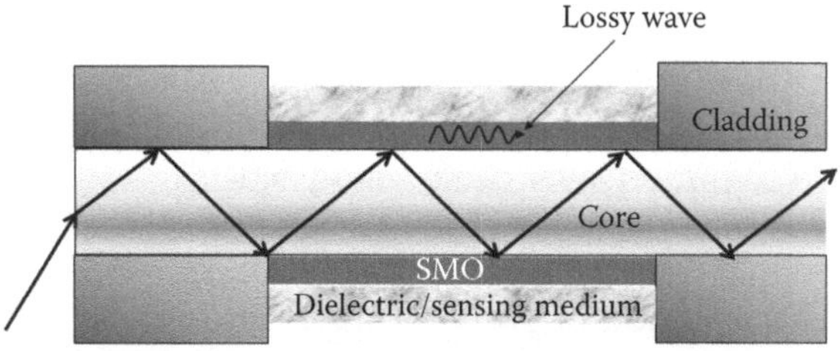

FIGURE 2.19
Fiber-based LMR configuration.

medium is taken as the calibration curve of the LMR-based fiber-optic RI sensor. A chemical or biosensor can also be realized using LMR technique. For this, a recognition layer is deposited over the SMO layer to interact with the sensing analyte/medium around it. The interaction between the recognition element and the sensing analyte changes the RI of the recognition element which in turn changes the resonance wavelength. This is the basic principle of the LMR-based chemical sensors.

The conditions of LMR can be discussed on the basis of the real and imaginary parts of the RI/dielectric constants of SMO, core of the fiber, and the surrounding medium. Let the dielectric constants of the optical substrate (fiber core) and SMO be $\varepsilon_{core} = \varepsilon_{r_{core}} + i\varepsilon_{i_{core}}$ and $\varepsilon_{SMO} = \varepsilon_{rSMO} + i\varepsilon_{iSMO}$, respectively, then the necessary conditions for LMR generation are assessed as (Del Villar et al. 2017):

1. SMO should possess a positive real part of its dielectric constant and its magnitude should be greater than its own imaginary part. Further, its value should be higher than that of the substrate and surrounding medium. Mathematically,

$$\begin{aligned}
&\varepsilon_{rSMO} > 0, \\
&\left|\varepsilon_{rSMO}\right| > \left|\varepsilon_{iSMO}\right|, \\
&\left|\varepsilon_{rSMO}\right| > \left|\varepsilon_{r_{core}}\right|, \\
&\left|\varepsilon_{rSMO}\right| > \left|\varepsilon_{s}\right|
\end{aligned} \tag{2.56}$$

2. If the RI of SMO and core are $n_{SMO} + ik_{SMO}$ and $n_{core} + ik_{core}$, respectively, then the imaginary part of RI of SMO should satisfy the relation

$$\begin{aligned}
&k_{SMO}^2 < n_{SMO}^2 - n_{core}^2, \\
&k_{SMO} < \left(\sqrt{2} - 1\right) n_{SMO}
\end{aligned} \tag{2.57}$$

When the guided mode propagating through the core of the optical fiber becomes lossy, due to the presence of SMO cladding, the propagation vector of the light propagating in the fiber changes. A change in the imaginary part of the dielectric constant changes the absorption properties of SMO. Similar to SPR-based sensors, by analyzing the shift in resonance wavelength and the change in the depth of the LMR spectrum, the real and imaginary parts of the RI of the recognition layer can be analyzed.

The SMOs that have been used for LMR sensing applications include indium tin oxide (ITO), zinc oxide (ZnO), indium oxide (In_2O_3), and tin oxide (SnO_2). The selection of SMO for application depends on the properties and compatibility of the material with the desired environment in which it has to work. The high conductivity of SMO material in the infrared region of electromagnetic spectrum supports sensor realization using SPR technique. In contrast, the optical transparency of the SMO material in the visible region of electromagnetic spectrum supports sensor based on LMR technique. The thickness of SMO material such as ITO also varies the conductivity of the material, which needs to be taken care of. A unique property of the LMR-based sensors is the generation of multiple LMR dips or a certain RI of the surrounding medium. Hence, by analyzing the maximum shift in resonance wavelength generated by each dip for a range of RI change, the best LMR dip can be chosen for the sensing purpose. The SMO-LMR-based sensors also possess high sensitivity due to the high value of the real part of the RI of SMO (Zamarreno et al. 2012). ZnO, one of the SMOs, exhibits dual property. One of the properties is supporting LMR and the other is the direct immobilization of biorecognition element by acting as a matrix which simplifies the fabrication process of biosensor probe (Usha et al. 2017). In addition, the chemisorption feasibilities on SMOs help in realizing a single SMO-layered LMR sensor probe as a high-sensitive gas sensor for environmental monitoring applications (Usha et al. 2015). Besides SMOs, polymers such as polyacrylic acid (PAA), polyallylamine hydrochloride (PAH), and polystyrene sulfonate (PSS) and their layer-by-layer combinations support LMR since their dielectric properties satisfy the conditions discussed above (Del Villar et al. 2012, Zamarreno et al. 2013). However, incompatibility of the polymers with certain viscous analytes limits their application in sensing.

In this section, a brief analysis is included for the readers to understand the interdependency of ZnO thickness and surrounding RI in generating the LMR spectrum. As discussed, for any material to exhibit LMR, the conditions (1) and (2) given by equations 2.56 and 2.57 need to be satisfied for the wavelength of operation chosen for the sensor characterization. Hence, the dispersion relation of the material needs to be plotted in the initial stage to analyze the dielectric property and its wavelength dependency. The Drude–Lorentz oscillator model for ZnO is given by

$$\varepsilon_2(\lambda) = \varepsilon_\infty - \frac{\omega_p^2}{\omega^2 + i\dfrac{\omega}{\tau}} + \frac{s_0\omega_0^2}{{\omega_0}^2 - \omega^2 + i\gamma\omega} \tag{2.58}$$

where the adequate parameters are $\varepsilon_\infty = 1$, $\omega_p = 4.28 \times 10^{15}$ rad/s, $s_0 = 1.9$, $\omega_0 = 10.02 \times 10^{15}$ rad/s, $\gamma = 3.103 \times 10^{14}$ rad/s, and electronic scattering time $(\tau) = 1.58 \times 10^{-15}$ s. The dispersion curve of ZnO is plotted in Figure 2.20.

The TE, TM, and TE + TM plots obtained using *N*-layer model discussed above for ZnO layer of 100 nm thickness coated over unclad core of 1 cm length are shown in Figure 2.21. The plots correspond to 1.33 RI of the surrounding medium, fiber core diameter 600 μm, and its NA as 0.37.

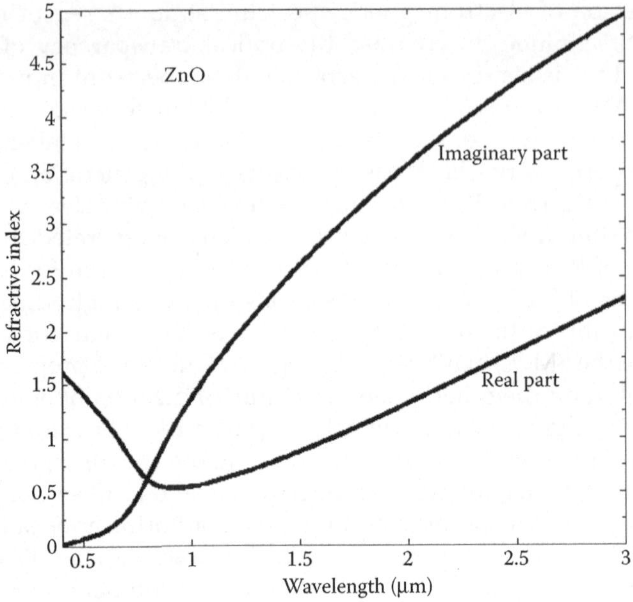

FIGURE 2.20
Dispersion curve of zinc oxide.

A single dip appearing in the visible region of electromagnetic spectrum at the wavelength of 422 nm in the TE plot (Figure 2.21a) upholds the dispersion relation of ZnO in the LMR range. In the case of TM plot (Figure 2.21b), two dips appeared, satisfying the LMR conditions in visible region at wavelength 411 nm and SPR condition at wavelength 1835 nm. The LMR spectrum for unpolarized light having both TE and TM modes possesses dips at 419 and 1828 nm showing both LMR and SPR, respectively, verifying the contributions of both TE and TM modes.

Choosing the TE + TM plot, the effect of thickness of ZnO on the LMR generation is analyzed for a surrounding RI of 1.33. Multiple LMR generation in ZnO with varying thicknesses is shown in Figure 2.22. For easy identification, LMR on the right side is named as LMR-1 (the first LMR) and the new LMRs generated on the left are named as LMR-2, LMR-3, and so on. The thickness of SMO to generate LMR can be chosen in such a way that the dip or dips appear at the desired wavelength or wavelength range suitable for the chosen application. Moreover, the sensitivity of each dip for a certain RI range will be different. Also, the range of RI of the sensing medium to which the LMR curve is sensitive varies with each LMR dip in the spectrum.

The thickness of SMO layer also affects the shift in resonance wavelength generated on each LMR dip in the spectrum. The behavior of LMR sensor with ZnO layer of thickness 50 nm for varying surrounding medium RI

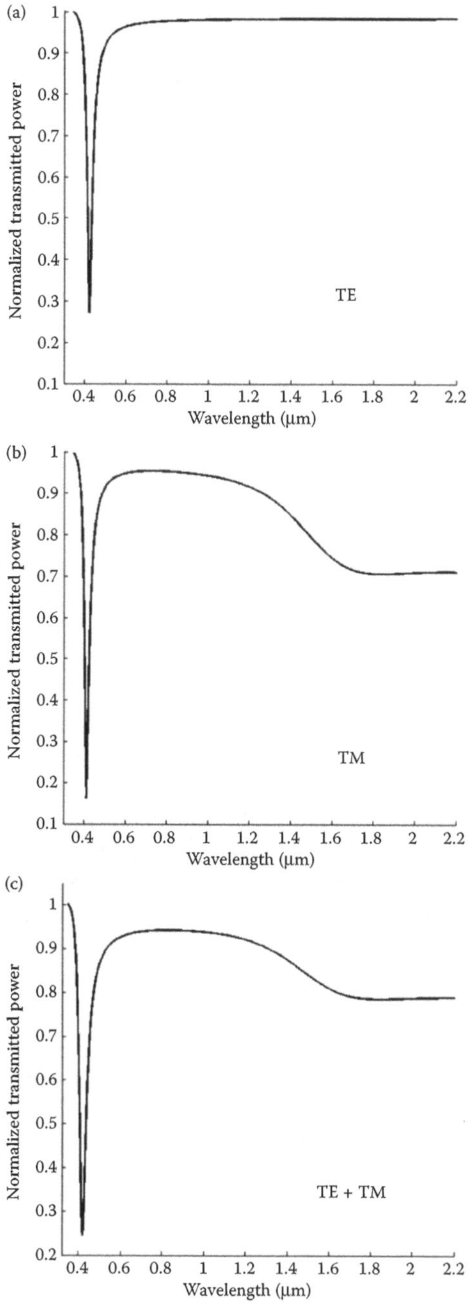

FIGURE 2.21
LMR generation for ZnO layer of 100 nm thickness coated over 1 cm unclad core region for (a) TE mode, (b) TM mode, and (c) TE + TM mode light propagation through the fiber core. The other parameters are: surrounding RI = 1.33, core diameter = 600 μm, and NA = 0.37.

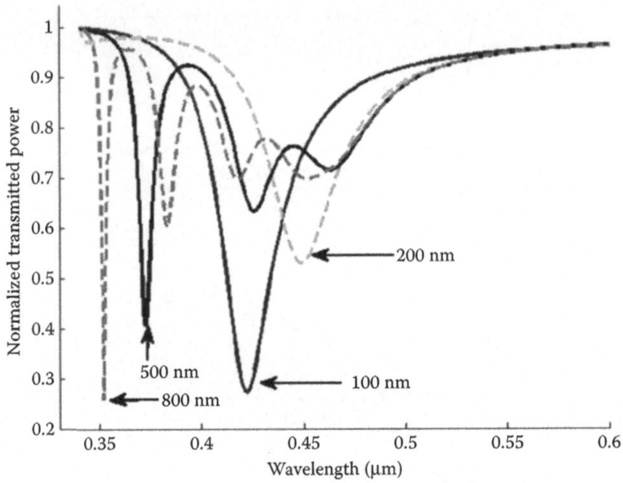

FIGURE 2.22
Multiple LMR generation with varying thickness of ZnO layer.

from 1.33 to 1.44 can be seen in Figure 2.23a. A maximum shift of around 28 nm in resonance wavelength occurs in LMR-1.

The shift can be different for different LMRs of the same spectrum. For understanding this, the LMR spectrum for the sensor with ZnO thickness of 400 nm is plotted for an RI change from 1.33 to 1.44. Three LMR dips are obtained as shown in Figure 2.23b and each dip has different shifts in resonance wavelength for the RI change from 1.33 to 1.44. The shifts in total resonance wavelength are 3 nm for LMR-1, 9 nm for LMR-2, and 2 nm for LMR-3. Thus, the plot signifies the fact that different LMR dips appear at different wavelengths in the electromagnetic spectrum and give different shifts for same change in RI. Hence, different shifts of different LMR dips of the same LMR spectrum allow us to pick the LMR dip of our choice. These unique properties of LMR make it the most approachable technique for sensing purposes. Similar to SPR technique, the changes in sensing length, core diameter, and NA do not cause any difference in the shift in resonance wavelength but change the depth of the LMR curve.

2.5 Summary

The chapter explicates the basics of resonance and how the technique of resonance can be used for sensing purposes. Among the RI-based transducer resonance mechanisms, which are commonly used for sensing applications,

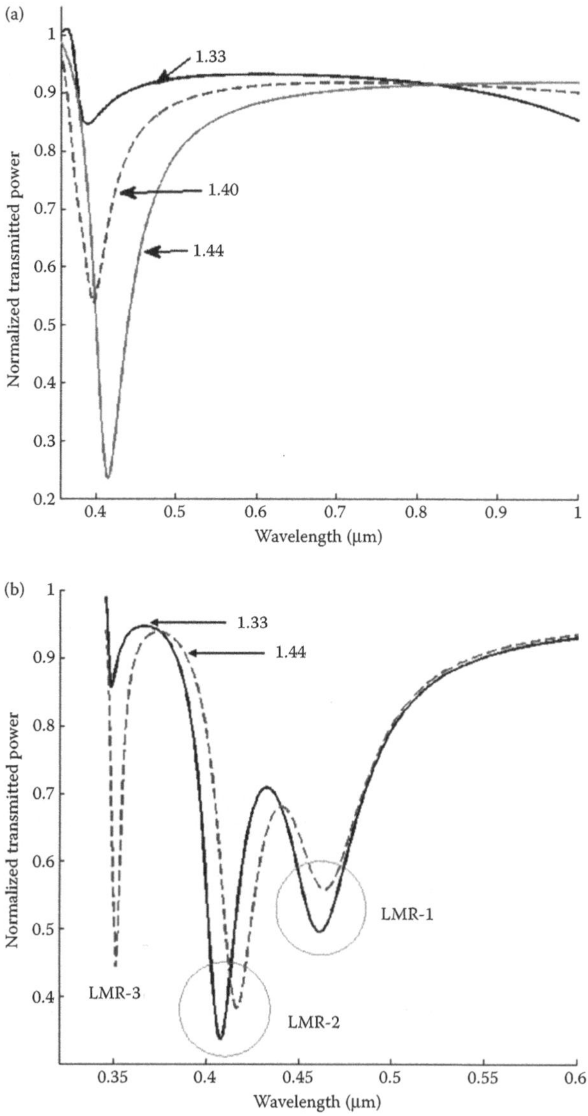

FIGURE 2.23
LMR spectra for different RI of sensing medium in the case of fiber-optic-based configuration with ZnO layer of (a) 50 nm and (b) 400 nm thickness coated over unclad core.

SPR, LSPR, and LMR have been given much attention. Basics and theory of SP and LSP have been discussed in detail. The conditions for SPR excitation, using the Otto and Kretschmann configurations, have been discussed along with various interrogation schemes. In addition to these, different types of fiber-optic resonances that can be combined with simple SPR for realizing high-sensitive sensors have been briefed. In the last section, the recently developed phenomenon of LMR using Kretschmann configuration on optical fiber substrate has been considered and the sensing performance has been analyzed by considering ZnO as the SMO material.

References

Berini, P. 2009. Long-range surface plasmon polaritons. *Adv. Opt. Photon.* 1: 484–588.

Bohren, C. F. and D. R. Huffman. 1998. *Absorption and Scattering of Light by Small Particles.* 2nd ed. New York: Wiley-Interscience Publications.

Del Villar, I., F. J. Arregui, C. R. Zamarreño et al. 2017. Optical sensors based on lossy-mode resonances. *Sens. Actuators B* 240: 174–185.

Del Villar, I., M. Hernaez, C. R. Zamarreno et al. 2012. Design rules for lossy mode resonance sensors. *Appl. Opt.* 51: 4298–4307.

Friedt, J. M. and L. A. Francis. 2016. Combined surface acoustic wave and surface plasmon resonance measurement of collagen and fibrinogen layer physical properties. *Sens. Biosens. Res.* 11: 60–70.

Gupta, B. D. and R. K. Verma. 2009. Surface plasmon resonance-based fiber optic sensors: Principle, probe designs, and some applications. *J. Sens.* 2009: 0979761.

Gupta, B. D., S. K. Srivastava, and R. Verma. 2015. *Fiber Optic Sensors Based on Plasmonics.* Singapore: World Scientific Publishing Company.

Homola, J., S. S. Yee, and G. Gauglitz. 1999. Surface plasmon resonance sensors: Review. *Sens. Actuators B* 54: 3–15.

Jorgenson, R. C. and S. S. Yee. 1993. A fiber-optic chemical sensor based on surface plasmon resonance. *Sens. Actuators B* 12: 213–220.

Kämpf, K., S. Kübler, F. W. Herberg, and A. Ehresmann. 2012. Magneto-optic surface plasmon resonance optimum layers: Simulations for biological relevant refractive index changes. *J. Appl. Phy.* 112: 034505.

Keathley, P. D. and J. T. Hastings. 2012. Nano-gap-enhanced surface plasmon resonance sensors. *Plasmonics* 7: 59–69.

Kretschmann, E. and H. Reather. 1968. Radiative decay of non-radiative surface plasmons excited by light. *Z. Naturforsch.* 23: 2135–2136.

Lahav, A., M. Auslender, and I. Abdulhalim. 2008. Sensitivity enhancement of guided-wave surface-plasmon resonance sensors. *Opt. Lett.* 33: 2539–2541.

Lee, B. H., Y. H. Kim, K. S. Park et al. 2012. Interferometric fiber optic sensors. *Sensors* 12: 2467–2486.

Liedberg, B., C. Nylander, and I. Lundström. 1983. Surface plasmons resonance for gas detection and biosensing. *Sens. Actuators* 4: 299–304.

MacPherson, W. N. 2014. Point sensors: Interferometric sensors. *Handbook of Optical Sensors.* eds. J. L. Santos, and F. Farahi. 351–376. Boca Raton, FL: CRC Press.

Maier, S. A. 2007. *Plasmonics: Fundamentals and Applications.* Germany: Springer Science & Business Media.

Marciniak, M., J. Grzegorzewski, and M. Szustakowski 1993. Analysis of lossy mode cut-off conditions in planar waveguides with semiconductor guiding layer. *IEE Proc. J. Optoelectron.* 140: 247–252.

Mayer, K. M. and J. H. Hafner. 2011. Localized surface plasmon resonance sensors. *Chem. Rev.* 111: 3828–3857.

Mie, G. 1908. Beiträge zur optik trüber medien, speziell kolloidaler metallösungen. *Annalen der physic*. 330: 377–445.

Nenninger, G. G., P. Tobiška, J. Homola, and S. S. Yee. 2001. Long-range surface plasmons for high-resolution surface plasmon resonance sensors. *Sens. Actuators B* 74: 145–151.

Otto, A. 1968. Excitation of nonradiative surface plasma waves in silver by the method of frustrated total reflection. *Z. Phys.* 216: 398–410.

Pile, D. F., T. Ogawa, D. K. Gramotnev et al. 2005. Two-dimensionally localized modes of a nanoscale gap plasmon waveguide. *Appl. Phys. Lett.* 87: 261114.

Pockrand, I., J. D. Swalen, J. G. Gordon, and M. R. Philpott. 1978. Surface plasmon spectroscopy of organic monolayer assemblies. *Surf. Sci.* 74: 237–244.

Powell, C. J. and J. B. Swan. 1960. Effect of oxidation on the characteristic loss spectra of aluminum and magnesium. *Phys. Rev.* 118: 640–643.

Ritchie, R. H. 1957. Plasma losses by fast electrons in thin films. *Phys. Rev.* 106: 874–881.

Shalabney, A. and I. Abdulhalim. 2010. Electromagnetic fields distribution in multilayer thin film structures and the origin of sensitivity enhancement in surface plasmon resonance sensors. *Sens. Actuators A* 159: 24–32.

Srivastava, S. K. and B. D. Gupta. 2013. Fiber optic plasmonic sensors: Past, present and future. *Open Opt. J.* 7: 1.

Turbadar, T. 1959. Complete absorption of light by thin metal films. *Proc. Phys. Soc. (London)* 73: 40–44.

Usha, S. P., A. M. Shrivstav, and B. D. Gupta. 2016. FO-SPR based dextrose sensor using Ag/ZnO Nanorods/GOx for insulinoma detection. *Biosens. Bioelectron.* 85: 986–995.

Usha, S. P., A. M. Shrivastav, and B. D. Gupta. 2017. Silver nanoparticle noduled ZnO nanowedge fetched novel FO-LMR based H_2O_2 biosensor: A twin regime sensor for in-vivo applications and H_2O_2 generation analysis from polyphenolic daily devouring beverages. *Sens. Actuators B* 241: 129–145.

Usha, S. P., S. K. Mishra, and B. D. Gupta. 2015. Fiber optic hydrogen sulfide gas sensors utilizing ZnO thin Film/ZnO nanoparticles: A comparison of surface plasmon resonance and lossy mode resonance. *Sens. Actuators B* 218: 196–204.

Wood, R. M. 1902. On a remarkable case of uneven distribution of light in a diffraction grating spectrum. *Proc. Phys. Soc. (London)*. 18: 269–275.

Zamarreno, C. R., P. Sanchez, M. Hernaez et al. 2012. Sensing properties of indium oxide coated optical fiber devices based on lossy mode resonances. *IEEE Sens. J.* 12: 151–155.

Zamarreno, C. R., P. Zubiate, M. Sagües, I. R. Matias, and F. J. Arregui. 2013. Experimental demonstration of lossy mode resonance generation for transverse-magnetic and transverse-electric polarizations. *Opt. Lett.* 38, 2481–2483.

Zheng, G., X. Zou, Y. Chen, L. Xu, and W. Rao. 2017. Fano resonance in graphene-MoS_2 heterostructure-based surface plasmon resonance biosensor and its potential applications. *Opt. Mat.* 66: 171–178.

3

Fiber-Optic Sensors

3.1 Optical Fiber as Sensor Element

In Chapter 2, we have introduced optical fiber as a substrate to define various types of resonances. In this chapter, before presenting application of optical fiber in sensing we would like to discuss it in more detail. Optical fiber, in general, consists of dielectric materials like glass and plastic. In a simplest form, it is a cylindrical structure consisting of a core with diameter of 5–600 μm surrounded by a cladding of slightly lower refractive index. It is further covered by a plastic coating to protect it from mechanical and environmental perturbations. A schematic of an optical fiber is shown in Figure 3.1. The light in the core propagates by means of total internal reflections (TIRs) at the core–cladding interface.

It has been more than four decades since studies on optical fiber and its applications in various areas unfolded. Most of the applications of optical fiber including communication and sensing were developed since the invention of lasers in 1960s. Although the prime motivation in the development was its feasibility in communication, later the research in optoelectronics, laser technology, and telecommunication areas paved the way for the advancement of optical fibers in sensing applications. The development of fiber from its high-loss medium to low loss in terms of transmission and attenuation laid a stepping stone to the modern world of fiber optics to support the telecommunication, electrical, civil, chemical, biological engineering fields, and industries to ease and better the work with finest features. The adoption of optical fiber for sensing applications resulted in the outset of a new product field which was supported by the research and developments in optoelectronics and laser technology which provided cost-effective components and improvements in its operation, size, and characteristics. With time, the demand of fiber-optic sensors has increased and it is still being considered the best choice for environmental monitoring and biomedical applications.

Optical fiber fits to be one of the best substrate choices for various applications because of its advantages including small size, flexibility, chemical inertness, cost-effectiveness, ability to withstand harsh environments, lightweight, safe operation in hazardous atmosphere, multidetection capability,

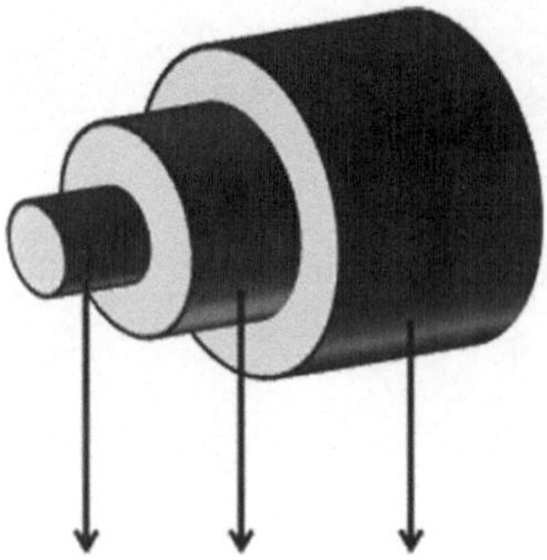

FIGURE 3.1
Schematic of an optical fiber.

immunity to electromagnetic interference, ability of *in vivo* application with online monitoring, and remote-sensing capabilities. Since the 1970s, the area of fiber-optic sensors has started evolving. In sensing applications, optical fiber is made sensitive to external variations which is evaluated by measuring the change in any or combination of the parameters of the light such as wavelength, frequency, phase, intensity, and polarization guided through the fiber. The design of the sensor probe, complexity in its fabrication, stability, and shelf life decide the performance, quality, and demand of the sensor. A simple design of sensor involves intensity modulation scheme, where a photodetector is used at the output end of a fiber, illuminated by a suitable optical source, which evaluates the variation in its output intensity with the change in external environment in which the senor is assembled (Culshaw and Kersey 2008). A suitably outlined fiber-based intrinsic sensor is capable to make measurements in terms of phase, polarization, and spectral modulation schemes. The variation in optical parameters is converted into a measurable signal. The important advancements in fiber-optic sensor field are the point of care and distributed design application capability (Busurin et al. 1985). For a fiber-based sensor system, three main components are optical source, detector, and the fiber. The fiber-optic point sensor with the sensing region at the end of fiber supports biomedical *in vivo* applications more easily than other methods of sensing such as electrochemical, acoustic, potentiometric, and so on. Fiber-based optical sensor works on any or combination of the spectrometric approaches/techniques such as luminescence, diffraction, scattering/absorptiometry, fluorescence, Faraday/Doppler effect, surface plasmon resonance (SPR), lossy mode resonance (LMR), interferometric, reflectometry/refractometry, etc.

In today's application, optical-fiber-based sensors play reliable role in aerospace, military applications, energy-producing systems, environmental parameter monitoring, and biomedical applications (Bogue 2011). Optical fiber sensors are used for online gas sensing as well with multipoint sensors.

These sensors permit remote sensing with the sensor head at one end of the fiber to analyze gases in great depths, detect the leakage of gases from gas pipelines, etc. The sensor is inserted through borehole to the required depth and analysis is performed online with the help of self-referenced arm which makes the evaluation easier without much complex calibrations (Bogue 2011). Fiber-optic sensors can also be used for the detection of leakage of oil, liquid benzene, and hexane due to their real-time monitoring capability. The environmental monitoring applications of optical fiber sensors also include water quality monitoring and soil standard evaluation for agricultural purposes. The industrial effluents, and by-products from human usage, as we know, end up in water bodies which are a considerable issue. In water bodies, the analysis is performed using fiber sensors for pesticides that result from their usage in agriculture, toxic waste products, and metal ions from chemical industries that affect the aquatic life and aquaculture. In agriculture, addition of pesticides affects the food quality as well as the standard of soil. Agricultural products and processed food items are reported with the presence of pesticides, bacteria, fungi, and pathogens. Fiber-optic sensors can be employed for monitoring the food quality as well. Soil water content/humidity (Cao et al. 2015), displacement (Iten et al. 2009), nutrients (Ramane et al. 2015), pH, and pathogens (Bosch et al. 2007) are measurable with the help of fiber-optic sensors, which perform simple evaluation and low-cost calibration with portable low-weight devices.

The approach of fiber optics in biomedical area is an already established and widely accepted one. The abilities of optical fiber having compatibility to surface restructuring, inertness to chemicals, biocompatibility, compliance with lasers, low-cost optoelectronic devices, and surgical instruments have made the fiber the finest choice of transducer in the medical field (Leung et al. 2007). The optical energy transmitted from source to the detector through optical fiber by TIR undergoes change in its transmission energy due to loss occurring in the presence of the medium in which the optical fiber is kept. The medium, which is inclusive of the sensing analyte, consists of a sensor layer designed accordingly for analyte detection. Using this concept, most of the fiber-optic sensors have been designed for practical applications. The choice of fiber, its structure, and parameters tailor the coupling of light from source to fiber and the characteristics of light passing through the fiber, which regulates sensing. Hence, a right choice of all the parameters is recommended for a perfect sensor which brings out its best performance.

3.2 Factors Affecting Light Propagation

A simple optical fiber is manufactured using silica glass which is amorphous in nature. The material is selected due to its inertness to chemical compounds and compatibility with biological samples, thus not affecting the physical and physiological environment unfavorably. The optical fiber minimizes the probability of contamination in the case of *in vivo* applications where the transfer of contagious diseases may occur between the patients. The buffer jacket usually given to the optical fiber having core and cladding as the fundamental parts ensures the protection of the fiber from adverse environment. The choice of the jacket depends on the application for which the fiber is opted. Teflon, polymer such as polyurethane, polyimide or PVC, and plastic are used as jacket layer (Roriz et al. 2013).

In the early days of implementing optical fiber for applications including communication, fibers were grouped into bundles to maximize the coupling of light. Later, with the invention of lasers, a single optical fiber started replacing the fiber bundle. The guidance of light through fiber is regulated by its various structural parameters such as core diameter, refractive indices of the core and cladding, etc. These structural parameters are used to define various important characteristic parameters of the fiber-like modes supported by the fiber, numerical aperture and the acceptance angle of the fiber, evanescent field in cladding medium, etc. We shall discuss these below.

3.2.1 Numerical Aperture and Acceptance Angle

In an optical fiber, light is transmitted by means of TIR. This happens when the angle of incidence of the light with the normal to the core–cladding interface is greater than or equal to the critical angle (θ_c).

Figure 3.2 schematically shows the propagation of rays in the core of the fiber. The critical angle of the fiber can be written as

$$\theta_c = \sin^{-1}\left[\frac{n_{clad}}{n_{core}}\right] \tag{3.1}$$

where n_{clad} and n_{core} are the refractive indices of the materials of cladding and core, respectively. The range of angle for which the condition of TIR

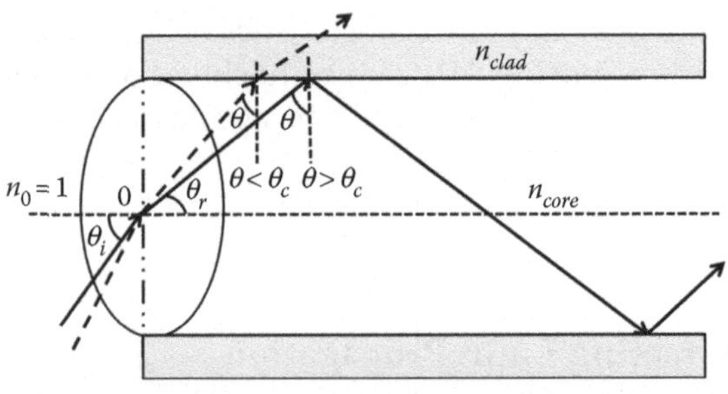

FIGURE 3.2
Light guidance in the fiber using total internal reflection phenomenon.

is satisfied for a light ray incident at the core–cladding interface varies from θ_c to 90°. Thus for the light guidance in the fiber, equation 3.1 can be written as

$$\sin\theta > \frac{n_{clad}}{n_{core}} \tag{3.2}$$

In Figure 3.2, we show the propagation of two rays with different angles with the normal to the core–cladding interface. It may be seen that one ray whose angle of incidence is greater than the critical angle gets propagated, while the other with smaller angle gets refracted through the cladding. For a ray incident onto the face of an optical fiber to get guided inside the core of the fiber, its angle with the axis of the fiber should be in the range of acceptance angle ($2\theta_A$) of the fiber, where θ_A is the half acceptance angle. The angle of acceptance of the fiber is related to its numerical aperture. Numerical aperture, denoted by *NA*, is the sine of the half acceptance angle, within which the light ray incident is guided along the core of the fiber. The numerical aperture of the fiber depends on the refractive indices of the materials of the core and the cladding. According to Snell's law of refraction,

$$n_0 \sin\theta_i = n_{core} \sin\theta_r \tag{3.3}$$

where n_0 is the refractive index of the medium outside the fiber and is equal to 1 in the case of air, θ_i is the angle of incidence of the ray onto the face of the fiber and θ_r is the corresponding angle of refraction. In terms of angle of incidence at the core–cladding interface (θ), Snell's law becomes

$$n_0 \sin\theta_i = n_{core} \cos\theta \tag{3.4}$$

As mentioned above, for the guidance of a ray inside the fiber, following condition should be satisfied: $\theta \geq \theta_c$. Using equation 3.1 and $n_0 = 1$ gives the following equation for the numerical aperture of the fiber

$$NA = \sin\theta_A = \sqrt{n_{core}^2 - n_{clad}^2} \tag{3.5}$$

The difference between the refractive index of the core and the cladding is generally around 1% for supporting the proper guidance of the light (Hecht 1997). The maximum angle of the propagating ray inside the core with the core–cladding boundary or the axis of the fiber is called as confinement angle (θ_{con}). Thus, *NA* can also be written as the sine of the confinement angle in free space. Mathematically,

$$NA = n_{core} \sin\theta_{con} \tag{3.6}$$

Thus,

$$\sin\theta_{con} = \frac{1}{n_{core}}\sqrt{n_{core}^2 - n_{clad}^2} \tag{3.7}$$

Therefore, for the propagation condition of light inside the core, the relation should be

$$\sin\theta_{con} \leq \sqrt{1 - \frac{n_{clad}^2}{n_{core}^2}} \tag{3.8}$$

The light ray making an angle greater than the confinement angle cannot propagate through the fiber. Using the above mentioned relations, the critical angle and the confinement angle for the light ray to undergo TIR for propagation through the core are obtained. However, the incident angle of the light ray to satisfy its coupling into the core for guidance is obtained from the following relation

$$n_0 \sin\theta_A = n_{core} \sin\theta_{con} \tag{3.9}$$

Therefore, for a ray incident on the face of the fiber with an angle of incidence, θ_i, with the axis of fiber to remain guided it should follow the following equation:

$$\sin\theta_i \leq \frac{1}{n_0}\sqrt{n_{core}^2 - n_{clad}^2} \tag{3.10}$$

Thus, the angle of incidence, confinement angle, and critical angle of the light ray for satisfying the propagating condition through the core of the optical fiber depend on the refractive index of the core as well as the cladding of the fiber and hence in turn on the numerical aperture. In all this, the numerical aperture decides the efficiency of coupling of light into the fiber and in turn influences the confinement and critical angle (Ghatak and Thyagarajan 1998).

3.2.2 *V*-Number and Fiber Modes

A continuous and steady propagation condition of light ray in an optical fiber is termed as a mode. It is the electromagnetic field distribution in a transverse plane that strengthens each other to give a steady state as it propagates along the fiber (Hecht 1997, Gupta et al. 2015). For a steady/standing wave to guide along a fiber, the light must follow the path of any of the mode supported by the fiber. The number of modes supported by a fiber is decided by its *V*-number. Different applications of fiber such as sensing and communications require different number of modes propagating and hence a different fiber too. For sensors using optical fiber as the substrate, the choice is generally for multimode or highly multimode fibers which provide higher space for interaction, strength, and durability than single mode fibers. The number of modes, a particular fiber can support for a particular incident light wavelength (λ), is evaluated by analyzing the parameters of fiber such as core diameter (D) and refractive indices of the core and cladding. The expression for the *V*-number is written as

$$V = \frac{\pi D}{\lambda}\left[\sqrt{n_{core}^2 - n_{clad}^2}\right] = \frac{\pi D}{\lambda} NA \tag{3.11}$$

Thus, the number of modes supported by the fiber depends on its numerical aperture as well. A fiber supports only a single mode if $V < 2.405$ and the number of propagating modes increases with the increase in *V*-value. A fiber is considered as highly multimode if $V > 10$. If $V > 1$, then the number of modes supported by the fiber in total is nearly $V^2/2$ (Gupta et al. 2015). A higher value of *V*-number causes loss in optical fiber.

3.2.3 Fiber Parameters

The core and cladding materials for fiber drawing are selected in such a way that the cladding confines/traps the light to the core of fiber (Subrahmanyam et al. 2012). Thus, the refractive index of the materials used for the fiber fabrication should ensure that the light be guided through the core with least loss of energy. For this, the refractive index of the core is always kept higher than that of the cladding which produces TIR of the electromagnetic ray hitting at the core–cladding boundary. The effective propagation of light ray through the fiber requires the relative refractive index deviation, $\Delta \ll 1$, where

$$\Delta = \frac{n_{core} - n_{clad}}{n_{core}} \tag{3.12}$$

In the case of step index fiber, there is an abrupt variation in the refractive index value from the core to the cladding while in the case of graded index fiber, the refractive index is maximum at the axis of the core and decreases continuously toward the core–cladding boundary with a constant refractive index of cladding. Fiber material can be glass, plastic, polymer, or a combination of these to follow the refractive index profile. In the case of a single mode fiber, the core diameter is very small, of the order of 8–15 μm, while for a multimode fiber core diameter varies over a broad range of 100–800 μm. As the core diameter increases, the number of modes supported by the fiber also increases thereby increasing the maximum light guidance. In fiber sensor probes where the sensing region can be made over clad or unclad fiber, the sensor performance depends on the number of reflections the light ray makes at the interface. In the case of SPR sensor, the core diameter affects the width of the SPR curve as well.

3.2.4 Evanescent Wave

As mentioned above, the light guidance inside the fiber is due to the TIR of the rays at the core–cladding interface. As we know, TIR occurs when a ray is incident from the denser medium at the interface of denser–rarer media. In the case of an optical fiber, the refractive index of the core is greater than that of the cladding and hence the TIR occurs. For the TIR, the angle of incidence of the ray should be greater than the critical angle, defined by equation 3.1. In TIR, light is not instantaneously reflected from the interface; instead of this, the superposition of incident and reflected beams occurs which forms a standing electromagnetic wave at the interface as shown in Figure 3.3a. The electric field amplitude is maximum at the interface and decays exponentially along the outward normal to the interface. The decaying field in the rarer medium is called evanescent field, while the electromagnetic wave associated with it is called an evanescent wave (Wang et al. 2016). The evanescent wave propagates along the denser–rarer media interface and its electric field amplitude (E) along the normal to the interface at a distance x is given by

$$E = E_0 \exp(-x/d_p) \tag{3.13}$$

where E_0 is the electric field amplitude at the interface and d_p, the penetration depth, is given by

$$d_p = \frac{\lambda}{2\pi(n_1^2 \sin^2\theta - n_2^2)^{1/2}} \tag{3.14}$$

where λ is the wavelength of light in vacuum and θ is the angle of incidence of the ray with the normal to the interface. Figure 3.3b shows the presence of evanescent field in an optical fiber. The field distribution in the core has its peak on the axis of the fiber. The evanescent field is extensively used for the chemical sensing. In the case of optical fiber, it is mostly used for the evanescent field absorption and plasmonic-based sensors. In the case of evanescent field absorption-based sensor, a part of the cladding is replaced by an absorbing medium as shown in Figure 3.4. The evanescent field is absorbed by the medium giving rise to a reduction in the power transmitted through the fiber. The absorption of evanescent field by the absorbing cladding medium has been exploited for the development of a large number of chemical and biological sensors. Larger the absorption, greater is

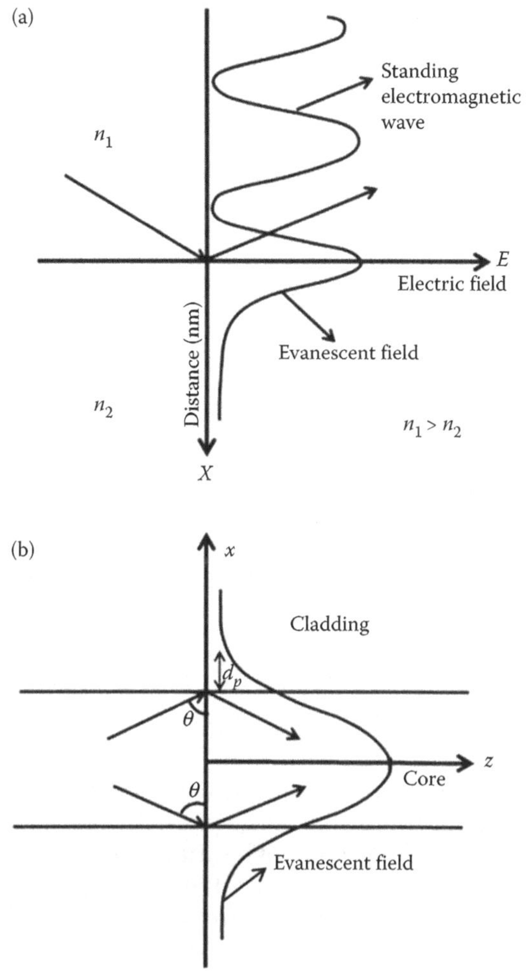

FIGURE 3.3
(a) Production of evanescent wave at an interface of denser–rarer media, and (b) evanescent field in cladding of an optical fiber. (Reprinted from *Sens. Actuators B*, 236, S. Pandita et al., In-situ monitoring of cross-linking reactions using E-glass fibers and evanescent wave spectroscopy, 358–366. Copyright 2016, with permission from Elsevier.)

the sensitivity of the sensor. For a given absorbing medium, the amount of absorption depends on the penetration depth of the evanescent field. An increase in the penetration depth increases the absorption and hence the sensitivity of the sensor. For a given angle of incidence, increase in the refractive index of the cladding medium increases the penetration depth. Further, for a given refractive index of the cladding medium, decrease in the angle of incidence at the core–cladding interface increases the penetration depth.

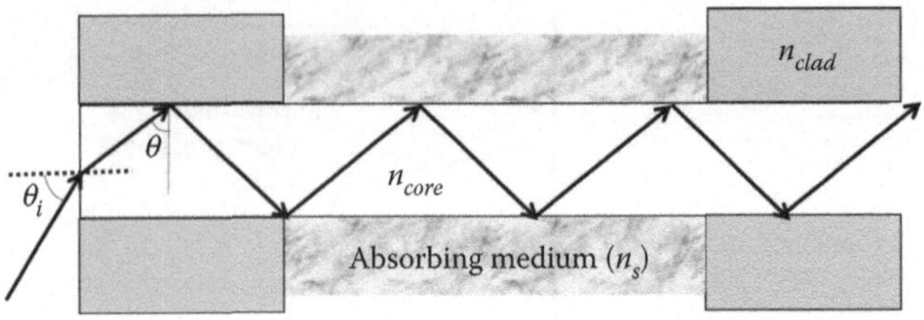

FIGURE 3.4
Schematic of an optical-fiber-based evanescent field absorption sensor.

3.3 Advantages of Fiber-Optic Sensors

The inherent advantages of optical fiber sensors over the conventional electronic sensors are the following:

Explosion proof: Since the primary signal in an optical fiber is optical, there will not be any risk of spark or fire in the case of sensors based on optical fiber. The advantage of this is that these sensors can be used in medical sciences without any risk to patients. Further, these can also be used in hazardous environments such as oil refineries, grain bins, mines, radioactive environments, and chemical processing plants.

Immunity to RF and EM interference: As the fibers consist of glass or plastics, the electrical isolation of patients is not required when the measurements are conducted using optical fiber sensors. Further, the functioning of the sensor will not be affected by the static electricity of the human body. In other words, the signals in optical fiber sensors cannot be interfered by the electromagnetic and radio frequency such as electrically noisy environment and strong magnetic field.

Environmental ruggedness and resistant: The permanent contact of the optical fibers with electrolyte solution and ionizing radiation cannot affect it because these are made from nonrusting materials such as plastics or glasses. Thus, they have excellent stability and can also withstand high temperature as high as 350°C.

Small size, lightweight, and flexible: The optical fibers have core diameter in the range 5–600 μm and therefore they can be bent round a mandrel. Further, a sensor with a very small probe can be fabricated over the fiber. Thus, the fiber-optic sensors can sense minute sample and can be used for invasive sensing in clinical chemistry and medicine. Due to their lightweight and small size, these sensors can be used in aircrafts.

Remote sensing: The optical fibers can also be used for remote sensing if low-loss fibers are used for probe fabrication. The remote sensing is important when the samples are hard to reach, dangerous, too hot or too cold, in harsh environments, or radioactive.

Potential of distributed sensing: The distributed sensing by optical fiber is one of the advantages where there is no competition. In distributed sensing, a physical or a chemical parameter/analyte can be measured as a function of position along the length of an optical fiber. These types of sensors find applications in industrial and environmental sensing.

Compactness: Using solid-state light source and detector, which are of small size and low weight, a compact optical fiber sensor can be designed.

3.4 Parameters Tailoring Sensor Performance

In order to evaluate the sensor, for a particular application, different parameters need to be figured out from the sensors response curve which qualify and quantify the characteristics and caliber of the sensor. In the case of fiber-optic sensor, the sensor's response would be in terms of transmittance power, reflectivity, refractivity, absorbance, and fluorescence emission. For instance, if the sensor is based on the absorbance (A) variation with respect to the change in concentration of the analyte, then the calibration/response curve can be plotted in terms of absorbance (A), relative absorbance (A/A_0) or relative change in absorbance $[(A - A_0)/A_0]$, where A_0 is the reference value of the absorbance taken for the standardization (D'Amico and Nitale 2001). Below we discuss important performance parameters of the sensor.

3.4.1 Sensitivity

One of the common evaluation parameters that analyze the sensor's performance irrespective of its design mechanism is sensitivity. It is, generally, defined as the quantitative change in the sensor's response signal for a measurable unit change in the quantity/property of the analyte. If the response of the sensor is analyzed in terms of absorbance with the change in analyte concentration, then the sensitivity can be evaluated by taking $\partial A/\partial C$, $\partial(A/A_0)/\partial C$ or $\partial[(A - A_0)/A_0]/\partial C$ depending on how the response curve of the sensor is plotted. Thus, predominantly, the sensitivity is calculated by taking the derivative of the output signal with respect to the input signal in the response curve. Few examples of feasibility of the output response nature of a sensor in terms of absorbance with respect to analyte concentration are shown in Figure 3.5a through f. The corresponding sensitivity plots

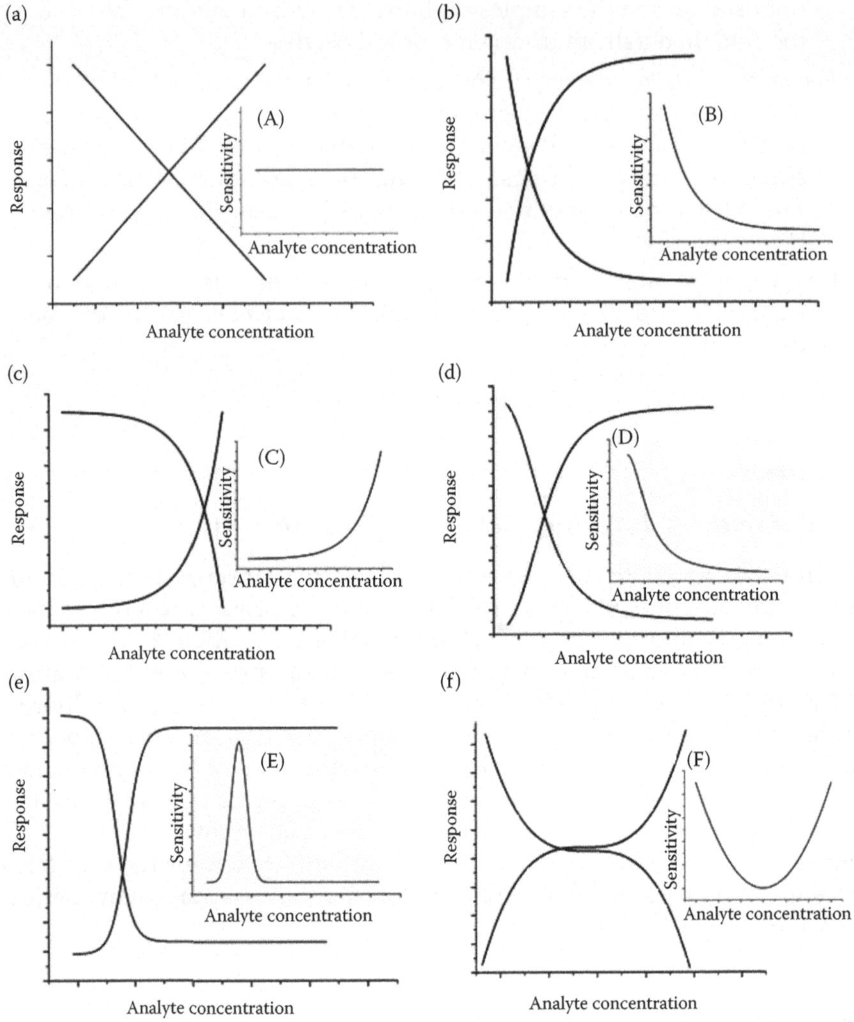

FIGURE 3.5
Various possible output response curves of the fiber optic sensor and their corresponding sensitivity taking absorbance as the output response and analyte as the concentration.

obtained by taking the derivative of the response curve are shown in Figures 3.5A through F. The variation can be linear or nonlinear. Here, for simple explanation and comparison, in all the examples, analyte concentration has been taken as the measurand and the absorbance as the response signal with $\partial A/\partial C$ as the sensitivity.

In the case of linear variation of the output signal with respect to the input measurand, the sensitivity remains constant for the entire range of operation of the sensor as shown in Figure 3.5a and (A). The linear response of the sensor is considered to be the best performing sensors, since it provides its maximum sensitivity for the complete range of operation (Kant et al. 2016). In the case of nonlinear variation of absorbance with respect to concentration, there can be many possibilities for the behavior of output signal with respect to input signal as shown in Figures 3.5b through f, where it can be polynomial, exponential, parabolic, logistic, sigmoidal, logarithmic, etc. Simple polynomial function fitting of the absorbance with analyte concentration is shown in Figure 3.5b with the sensitivity plot behavior shown in 3.5(B) (Shrivastav and Gupta 2016). For both the cases of 3.5(b), whether the response signal is increasingly/decreasingly saturating, the nature of sensitivity curve will be the same with higher sensitivity for lower concentrations. Figure 3.5c and (C) are the response and sensitivity plots for exponential change of absorbance with concentration, where the sensitivity also follows the same behavior (Jorgenson and Yee 1993). The logistic variations in the response curve are shown in Figure 3.5d and e. For both the cases, the sensitivity curve has a peaked behavior as shown in Figure 3.5D and E (Lagakos et al. 1987). A plot of sensitivity with parabolic behavior is shown in Figure 3.5F for the response curve shown in Figure 3.5(f) (Singh and Gupta 2012).

3.4.2 Selectivity

The ability of a sensor to be selective for a particular analyte is called the selectivity. The ratio or percentage of selectivity shown for a desired measurand/analyte by the sensor in the presence of other measurands is also classified under selectivity. The effect of other measurands present in the analyte on the sensing of a particular analyte/measurand is called cross-selectivity. This exists when there are multiple physical/chemical/biological parameters that can act as measurands in a particular analyte (D'Amico and Nitale 2001).

Intensification or enhancement of selectivity is called specificity. It denotes the degree of accuracy to which the desired analyte can be sensed in a mixture or real sample (Danzer 2001). The sensor extremely selective for a particular analyte is considered as specific (Vessman 1996). Selectivity is a parameter that can be graded or calibrated accordingly whereas specificity is a definite parameter (Boef and Hulanicki 1983). The hindrance offered by other analytes in a mixture or real sample, affecting the sensing of the desired analyte in the sample, is called interference (Boef and Hulanicki 1983, Vessman 1996, Danzer 2001). For industrializing a sensor, the sensor probe should have results in good agreement in terms of its interference study. The terms selectivity and specificity are also used in analysis of chemical/biological samples for multiple analyte and single-analyte analysis, respectively (Danzer 2001). However, the conditions and definition of rendering selectivity and sensitivity differ depending on the experimental and environmental conditions, analyte, the sample/mixture/real sample in which the analyte has to be sensed, the method of analysis, and the quantitative or qualitative determination (Vessman 1996, Danzer 2001).

3.4.3 Limit of Detection

The lowest concentration of the target molecule in a solution that is detected by means of an analytical method is generally termed as the limit of detection (LOD). Although its definition and concept is simple, there is always an uncertainty in the calculation of its value depending on the technique, calibration, and the noises and errors involved in the experiment. Hence, the International Union of Pure and Applied Chemistry (IUPAC) in 1975 proposed a concept for the calculation of detection limit as, "The limit of detection, expressed as a concentration (LOD) is derived from the smallest measure that can be detected with reasonably certainly for a given analytical procedure." This concept was again reframed by ACS Subcommittee on Environmental Analytical Chemistry in 1980 as, "Limit of detection is the lowest concentration of analyte that the detection scheme can detect reliably." Different definitions/methods have been used for the detection limit by the researchers. These are named as the method detection limit (MDL), instrument detection limit (IDL), practical quantification limit (PQL), and limit of quantification (LOQ) (Long and Winefordner 1983).

The most common method for evaluating the LOD of a sensor is based on the signal-to-noise ratio of the device. In this method, the detection limit is calculated by taking a multiplication factor of 3 with the standard deviation near the blank concentration (σ) (Shrivastava and Gupta 2011).

$$\mathrm{LOD} = 3 * \sigma \tag{3.15}$$

For the sensing device having linear response of calibration curve, the instrument response (denoted by y) is linearly related to the analyte concentration (x) and is expressed as

$$y = \alpha + \beta x \tag{3.16}$$

where α and β are constants. According to this, LOD of the sensor is expressed as

$$\mathrm{LOD} = 3\left(\frac{S_y}{\beta}\right) \tag{3.17}$$

where S_y is the standard deviation calculated in the sensor response, while β denotes the slope of the calibration curve. There are few other cases where LOD of the sensor is calculated by taking the ratio of the instrumental resolution and the sensitivity of the sensor near-blank concentration. This is expressed as (Gupta et al. 2016)

$$\mathrm{LOD} = \left(\frac{\Delta y}{S_{blank}}\right) \tag{3.18}$$

where Δy is the resolution of the detecting and analyzing instrument such as spectrometer and S_{blank} is the sensitivity of the sensor near-blank concentration.

3.4.4 Limit of Quantification

The LOQ is defined as the lowest analyte concentration which is possible to be both sensed and quantified by the sensor within the defined limit of certainty after repeating the experiments. It is also important to note that LOD and LOQ of the sensor are rarely the same but usually the value of LOQ is higher than LOD (Long and Winefordner 1983). Generally, LOQ of a sensor is evaluated as 10 times the standard deviation near the zero concentration of analyte and is expressed as (Shrivastava and Gupta 2011)

$$\mathrm{LOQ} = 10 * \sigma \tag{3.19}$$

In the case of linear calibration curve of sensor, LOQ can be estimated as

$$\mathrm{LOQ} = 10\left(\frac{S_y}{\beta}\right) \tag{3.20}$$

where S_y stands for the standard deviation in the sensor response, while β denotes the slope of calibration curve. In some studies, quantification limit is calculated as the ratio of the standard deviation of the response to the sensor sensitivity near-blank concentration of analyte and is expressed as (Agrawal et al. 2016)

$$\mathrm{LOQ} = \frac{\Delta y_{blank}}{S_{blank}} \tag{3.21}$$

where Δy_{blank} is the standard deviation of the output signal and S_{blank} denotes the sensitivity of sensor near-blank concentration of analyte.

3.4.5 Repeatability

The degree of probability to which the values are repeated in the experiment is checked by repeatability study. The same sensor parameter needs to be repeatedly measured, and the variations in the measurement show whether the sensor satisfies repeatability or not. For example, at a particular analyte concentration, the absorbance parameter needs to be measured. The sensor is experimented repeatedly for the selected concentration, keeping the working conditions fixed. Thus, in mathematical terms, repeatability is defined as the changeability in the median value when the same experiment is repeated by the same operator. This also shows the preciseness and definitude of the sensor system chosen for a particular application. The quality of data/response obtained from a sensor depends on the accuracy to which the measurement is repeated by performing the experiments multiple times.

3.4.6 Reproducibility

The parameter reproducibility is defined as the changeability in the median value when the same experiment is repeated by different operators. It qualifies and quantifies a system by checking the experimental measurements repeated over a long period of time. Reproducibility thus incorporates repeatability. The percentage of reproducibility is evaluated in simple terms by taking the ratio of the obtained result and expected result, multiplied by 100. Stability parameter of sensor is interrelated with the repeatability and reproducibility assessment factors. For example, by measuring the absorbance of a particular analyte concentration by keeping the sensor in the analyte vicinity continuously shows the stability behavior of the sensor system. By performing the stability experiments multiple times, the repeatability and reproducibility parameters of the sensor can also be evaluated.

3.4.7 Detection Accuracy

The experimental precision that can be attained by the sensor is termed as detection accuracy. There are many factors that affect sensor's performance and the results generated throughout its operation, such as the environmental noise, noise generated by mechanical machines, and electronic components used in the sensor design and human error. The percentage of unpredictability in a measurement in comparison to the standard results inclusive of the error values is the detection accuracy. In other words, the deviation of the measurement value with respect to the standard value determines the accuracy and inaccuracy.

3.4.8 Figure of Merit

Figure of merit (FOM) of a sensor is the performance parameter which measures the efficiency and effectiveness of the system. Generally, it is calculated, for example, in the case of plasmonic sensor, by taking the ratio of sensitivity and full width at half maximum (FWHM) of the spectrum of the sensing device. In the case of high FWHM, detection accuracy is low as mentioned above. Thus, for a sensor with low-sensitivity and low-detection accuracy, FOM is low and the sensor is not able to detect the analyte precisely. However, if the sensor has better detection accuracy (low FWHM) and low sensitivity, then also it limits the sensor's realization. Hence, a combinational ratio of the parameters, sensitivity, and the detection accuracy (inverse of FWHM) is important for a sensor (Gupta et al. 2015).

3.5 Designs of a Fiber-Optic Sensor Probe

Devices with high-performance capability or improved detection performance are the most demanded deed in the field of sensing. There are different ways of improving sensor's performance: designing the combination by choosing the suitable transducer, bio-recognition element, matrix (optional), substrate with supportive material, and structural geometry. Most of the times, the bio-recognition element, matrix, and their combination, which form the sensing layer, are varied to improve sensor's performance which intensifies the probe fabrication procedure and in turn affects its storage and long-term stability conditions. This is a major drawback happening with biological sensors. In the case of a sensor designed using optical fiber substrate, the performance of sensor can be easily improved by modifying the structural design/geometry of optical fiber. The optical fiber allows its geometry to be either straight, U-shaped, tapered, or D-shaped to improve the probe performance. In this section, first we shall discuss the simplest straight fiber-optic probe and then modify it to obtain the better performance of the sensor.

3.5.1 Straight Probe

In the case of a sensor implemented with fiber-optic substrate, the sensing region is generally made by uncladding a particular length from the middle region of a long fiber. The schematic of the probe is shown in Figure 3.4. The length of the sensing region is chosen according to the design or requirement of the probe. A very basic fiber-optic sensor works on the evanescent field absorption by the analyte around the unclad region of the fiber mentioned in Section 3.2.4. The absorption of evanescent field depends on the length of the unclad region and the number of reflections of a ray in this region. The larger is the number of reflections, the greater is the absorption. If P_0 is the power transmitted by the fiber in the absence of absorbing medium around the core, then the power transmitted in the presence of absorbing medium is given by

$$P = P_0 \exp(-\gamma L) \tag{3.22}$$

where L is the length of the unclad region of the fiber with sensing medium around it and γ is the evanescent absorption coefficient of the medium which is written as

$$\gamma = NT \tag{3.23}$$

where T is the Fresnel transmission coefficient of the light at the interface of a lossless core and lossy cladding and N is the number of reflections per unit length of a ray with angle of incidence θ at the core–cladding interface in the sensing region. Further, T increases if the penetration depth of the evanescent field increases. If n_s -ik is the refractive index of the sensing medium, then T can be written as (Gupta and Singh 1994a,b)

$$T = \frac{\alpha \lambda n_s \cos\theta}{\pi(n_{core}^2 - n_s^2)(n_{core}^2 \sin^2\theta - n_s^2)^{1/2}} \tag{3.24}$$

where $\alpha = 4\pi k/\lambda$ is the bulk absorption coefficient of the sensing medium at wavelength λ. For a core of radius ρ, the number of reflections per unit length in the sensing region is given by

$$N = \frac{\cot\theta}{2\rho} \tag{3.25}$$

The evanescent absorption coefficient for a given angle θ and the refractive index of the sensing medium (n_s) is, thus, given by

$$\gamma(\theta, n_s) = \frac{\alpha \lambda n_s \cos\theta \cot\theta}{2\rho\pi(n_{core}^2 - n_s^2)(n_{core}^2 \sin^2\theta - n_s^2)^{1/2}} \tag{3.26}$$

If all the guided rays are launched into the fiber, then the effective evanescent absorption coefficient of the sensing medium is given by (Gupta et al. 1993)

$$\gamma_{eff}(n_s) = \frac{\int_{\theta_c}^{\pi/2} P(\theta)\gamma(\theta, n_s)d\theta}{\int_{\theta_c}^{\pi/2} P(\theta)d\theta} \tag{3.27}$$

where $P(\theta)$ is the power distribution of the rays.

Consider the launching of light in the fiber using a collimated source focused onto the face of the fiber at its axis using a microscope objective as shown in Figure 3.6. If we consider the uniform distribution of the power across the plane of the beam, then to find the power distribution of the ray incident on the face of the fiber, we consider an annular ring of angular radius θ_i and $\theta_i + d\theta_i$; θ_i is the angle of the ray from the axis for the ray outside the fiber. The area of the ring is equal to the circumference of the ring multiplied by the width of the ring. The circumference of the ring is $2\pi f \tan\theta_i$, while the thickness of the ring is $d(f\tan\theta_i) = fd\theta_i/\cos^2\theta_i$. The area of the ring thus becomes

$$Area = \frac{2\pi f^2 \tan\theta_i}{\cos^2\theta_i} d\theta_i \tag{3.28}$$

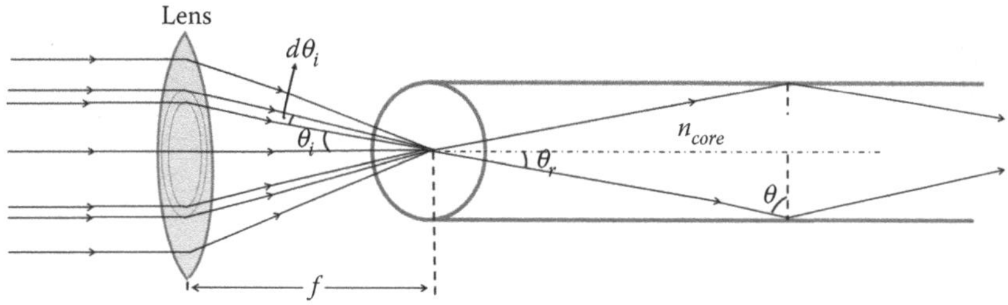

FIGURE 3.6
Light launching in an optical fiber from a collimated source using a microscope objective.

The power dP arriving at the axial point of the fiber from the lens between angles θ_i and $\theta_i + d\theta_i$ can then be written as

$$dP \propto \left(\frac{\tan\theta_i}{\cos^2\theta_i}\right) d\theta_i \tag{3.29}$$

The ray incident on the face of the fiber gets refracted and then reflected due to TIR from the interface of the core–cladding as shown in the figure. If θ_r is the angle of refraction of the ray inside the fiber, then the angle of incidence at the core–cladding interface is written as

$$\theta = 90° - \theta_r \tag{3.30}$$

Using Snell's law, it gives

$$\sin\theta_i = n_{core}\sin\theta_r = n_{core}\cos\theta \tag{3.31}$$

Differentiation of equation 3.31 gives

$$d\theta_i = \frac{n_{core}\sin\theta}{\sqrt{1 - n_{core}^2\cos^2\theta}} d\theta \tag{3.32}$$

Using equations 3.31 and 3.32, the power distribution is

$$dP \propto \frac{n_{core}^2\sin\theta\cos\theta}{\left(1 - n_{core}^2\cos^2\theta\right)^2} d\theta \tag{3.33}$$

Substitution of equation 3.33 in equation 3.27 gives the effective evanescent absorption coefficient of the sensing medium when the light is launched using a collimated beam focused on the axial point on the face of the fiber using a microscope objective.

If the light is launched in the fiber using a diffuse or the Lambertian source like light-emitting diode (LED) placed against the fiber end face covering at least the cross section of the core, then the power dP at the end face of the fiber between angles θ_i and $\theta_i + d\theta_i$ is proportional to $\sin\theta_i \cos\theta_i d\theta_i$. In terms of angle θ, it can be written as (Gupta and Singh 1994a,b)

$$dP \propto n_{core}^2 \sin\theta \cos\theta d\theta \tag{3.34}$$

The evanescent field absorption sensors are, generally, used to monitor the concentration of an absorbing fluid. The sensitivity of such sensors is defined as the fractional change in power per unit change in the concentration of the sensing medium. Mathematically, it can be written as

$$S = -\frac{1}{P(L)}\frac{dP(L)}{dc} \tag{3.35}$$

where L is the length of the sensing region, c is the concentration of the sensing medium, and P is the power transmitted by the fiber. The sensitivity has been shown to be directly proportional to γ/α (Gupta et al. 1993). Thus, an increase in the evanescent field absorption coefficient increases the sensitivity of the sensor.

In Chapter 2, we have discussed SPR-based fiber-optic sensor. In SPR-based sensor, the unclad core is coated with a thin metal layer and the dielectric medium to be sensed is kept around it. The principle of SPR-based sensor is different from the evanescent field absorption-based sensor. Evanescent field absorption sensors are simply based on the absorption of evanescent field by the absorbing sensing medium, while the SPR-based sensors are based on the transfer of energy of the TM-polarized evanescent wave to surface plasmon wave resulting in a dip in the transmitted spectrum. Therefore, instead of recording absorbance, the wavelength corresponding to minimum transmission is recorded in the case of SPR-based sensors. In the case of sensors based on lossy mode resonance, instead of metal layer over unclad core, a film of semiconductor metal oxide is coated. A slight modification is made in equation 3.27 for these two types of sensors as can be seen in Chapter 2. The sensors based on SPR and LMR have been discussed in Chapter 2.

3.5.2 U-Shaped Probe

As mentioned above, to enhance the sensitivity of the sensor the geometry of the unclad region is modified. One of the geometries chosen is a U-shaped which was first used for the evanescent field absorption sensor (Gupta et al. 1996). In U-shaped regime, the number of reflections occurring at the core-medium interface and the evanescent field penetration increases from that of a straight probe discussed in previous section. Thus, in the

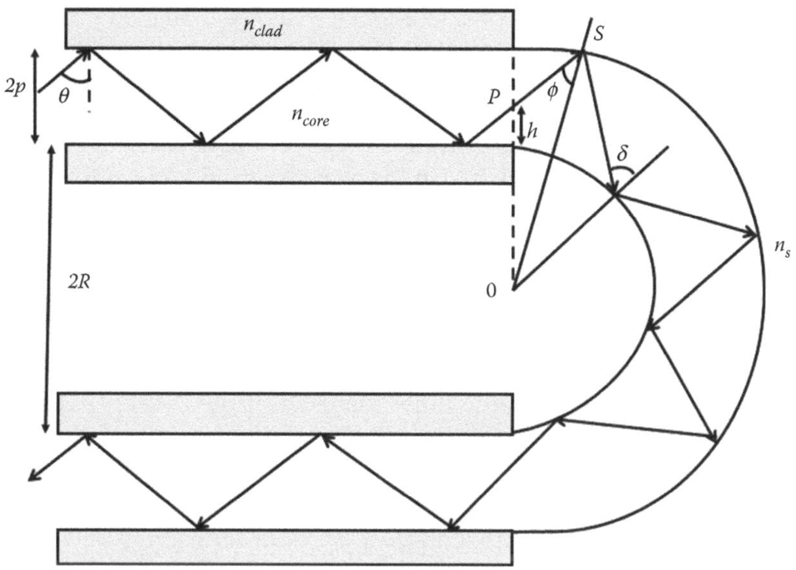

FIGURE 3.7
Schematic of a U-shaped fiber probe showing the light guidance by reflection from the outer and inner surfaces of fiber core-sensing medium interface.

case of U-shaped fiber-optic probe with absorbing sensing medium around the unclad region, the transmitted power through the fiber is expected to decrease more. Figure 3.7 shows a schematic of fiber-optic U-shaped probe. The unclad region is replaced by the sensing medium and hence the angle at which the TIR takes place from the fiber core-sensing medium interface depends on the refractive index of the sensing medium (Takeo and Hattori 1982, 1992). Since the refractive index of the sensing medium is, generally, smaller than the refractive index of the cladding of the fiber, the critical angle of the sensing region is less than that of the fiber. Therefore, in equation 3.26, the minimum value of angle θ is the critical angle of fiber and it cannot be equal to the critical angle of the sensing medium. Thus, one cannot achieve large penetration depth of the evanescent field and hence the absorption will be small which means the sensitivity of the sensor will be poor. The U-shaped probe enhances the penetration depth and hence the sensitivity of the sensor. In the U-shaped region, the angle of incidence of the ray decreases as the ray enters from the straight region. Due to this, the angle θ of the ray approaches the critical angle of the sensing bent region, and hence the penetration depth and the evanescent absorption increase. As the bending radius of the U-shaped region decreases, the angle of incidence approaches the critical angle of the sensing region. Thus, one can expect an increase in the sensitivity of the sensor.

To analyze the U-shaped probe and compare its performance with straight probe, a 2D approach and the propagation of meridional rays confined in the plane of bending have been considered. In the case of U-shaped probe, there can be two types of ray propagation. One, a ray gets totally reflected both at the outer and inner surfaces and the second, the ray gets totally reflected from outer surface and does not reach the inner surface in the bent region. The angle of incidence, θ, in the straight region of the fiber changes to angle of incidence, ϕ, at the outer surface in the bent region and relates to the angle θ according to the following expression:

$$\sin\phi = \left[\frac{R+h}{R+2\rho}\right]\sin\theta \tag{3.36}$$

where h is the height at which the light ray incidents at the entrance of the core-unclad region and R is the inner diameter of the U-shaped structure formed by the bending of the fiber. In a similar way, one can write relation between angle of incidence at the inner surface of the bent region, δ, and the angle θ as

$$\sin\delta = \left[\frac{R+h}{R}\right]\sin\theta \tag{3.37}$$

For the rays lying in the plane perpendicular to the plane of bending, the angle of the rays in the straight portion of the core and in the U-shaped portion will remain the same. In other words, there will not be any change in the angle of the ray as it propagates through the fiber. Thus, the expression for the effective evanescent absorption coefficient for such rays will be the same as obtained for straight probe. Thus, the total effective evanescent absorption coefficient for U-shaped probe is given as

$$\gamma_{eff-Tot} = \frac{1}{4}\left|\gamma_{eff}(\phi_1,\phi_2)\right|_{outer} + \frac{1}{4}\left|\gamma_{eff}(\delta_1,\delta_2)\right|_{inner} + \frac{1}{2}\gamma_{eff}(\theta_1,\theta_2) \tag{3.38}$$

where the range of inner angles varies from ϕ_1 to ϕ_2 and outer angles vary from δ_1 to δ_2. The angles $\theta_1 = \sin^{-1}(n_{clad}/n_{core})$ and $\theta_2 = \pi/2$. The angles ϕ and δ are determined using these values of θ_s. The effective evanescent absorption coefficient for the outer surface of the sensing U-shaped probe is written as

$$\left|\gamma_{eff}(\phi_1,\phi_2)\right|_{outer} = \frac{\displaystyle\int_0^{2\rho}\int_{\phi_1}^{\phi_2} P(\theta)\gamma(\theta)d\theta dh}{\displaystyle\int_0^{2\rho}\int_{\phi_1}^{\phi_2} P(\theta)d\theta dh} \tag{3.39}$$

The effective evanescent absorption coefficient for the inner surface can be obtained in the same way.

Considering the factors affecting the sensitivity of the U-shaped probe, it has been reported that the sensing length and bending radius of the fiber probe are important. The sensitivity increases initially with a decrease in the bending radius of the sensing region up to a particular bending radius (R), while a further decrease in R causes a decrease in the sensitivity. This is because the incident angle of the final ray becomes less than that of the guided ray (to satisfy the condition of TIR), which in turn decreases the sensitivity (Khijwania and Gupta 2000). As the sensing length increases, the number of reflections increases, that finally increases the absorption of the evanescent field in the sensing medium. Thus, as the sensing length increases, the sensitivity of the sensor increases. SPR and localized SPR-based fiber-optic sensors have been reported in the literature where the use of U-shaped probe has been made. We shall discuss these sensors in the next chapters.

3.5.3 Tapered Probe

The another method of increasing the penetration depth or evanescent field absorption and hence the sensitivity of the evanescent field absorption sensors is changing the shape of straight region of the probe to taper. In a tapered probe, the core diameter is smoothly reduced along the length of the fiber. A bitapered fiber probe has a down-tapered as well as an up-tapered regime, where the core diameter recedes in the down-tapered region, while the diameter expanses in the up-tapered region as shown in Figure 3.8.

The figure shows the unclad tapered core as the sensing regime having a sensing length of $2L$. As the modes propagate in a multimode fiber through its tapered regime, due to the decrement in the core diameter, all the modes could not propagate. This results in a loss of some of the modes and their evanescent fields. In terms of ray optics, the tapered region offers a decrease in the angle of incidence and as the ray propagates its angle with the normal to the core-sensing medium, interface approaches critical angle of the sensing region. As a result of this, the evanescent field in the sensing medium is more as compared to straight probe. Moreover, the

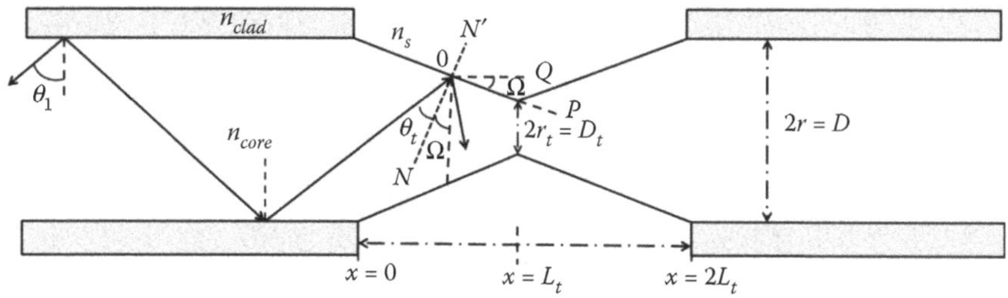

FIGURE 3.8
Schematic of a bitapered fiber optic probe for enhancing the sensitivity of the fiber optic sensor.

decrease in the core diameter increases the number of reflections, which further increases the absorption of the evanescent field in the sensing medium. Thus, the tapered probe increases the sensitivity of the sensor as compared to the usual fiber-optic straight probe (Gupta et al. 1993, Verma et al. 2008, Hartung and Bartelt 2011, Kerttula et al. 2012, Sony and Soumya 2013, Gupta et al. 2015).

Consider the propagation of a guided ray in the fiber through the tapered region. As can be seen from Figure 3.8, in the tapered region, the angle of the incidence of the ray (θ_t) is different from that the ray makes in the straight region (θ). As the ray propagates, it decreases along the length of the taper and follows the following relation:

$$\theta_t(x) = \cos^{-1}\left[\frac{\rho\cos\theta}{\rho(x)}\right] - \Omega \tag{3.40}$$

where $\rho(x)$ is the core radius as a function of distance x from the input end of the taper along its length and Ω is the taper angle given as

$$\Omega = \tan^{-1}\left(\frac{\rho - \rho_0}{L}\right) \tag{3.41}$$

where ρ_0 is core radius at the output end ($x = L$) of the taper. Thus, $\theta_t(x)$ depends on $\rho(x)$ as shown in the above expression. It decreases as $\rho(x)$ decreases. Hence, for a particular ray to remain guided in the tapered region of the core, the condition $\theta_t(x) \geq \sin^{-1}(n_s/n_{core})$ should be satisfied. To analyze the tapered probe, following three types of profiles have been used (Gupta and Singh 1994b)

Linear:

$$\rho_l(x) = \rho - \frac{x}{L}(\rho - \rho_0) \tag{3.42}$$

Parabolic:

$$\rho_p(x) = \left[\rho^2 - \frac{x}{L}(\rho^2 - \rho_0^2)\right]^{1/2} \tag{3.43}$$

Exponential linear:

$$\rho_{el(x)} = (\rho - \rho_0)\left[e^{\left(-\frac{x}{L}\right)} - \frac{x}{L}e^{-1}\right] + \rho_0 \tag{3.44}$$

The effective evanescent absorption coefficient in the case of tapered probe is thus given by

$$\gamma_{eff}(\rho_0) = \frac{\int_0^L dx \int_{\theta_{t1}(x)}^{\theta_{t2}(x)} P(\theta)\gamma(\theta, x)d\theta}{\int_0^L dx \int_{\theta_{t1}(x)}^{\theta_{t2}(x)} P(\theta)d\theta} \tag{3.45}$$

The angles $\theta_{t1}(x)$ and $\theta_{t2}(x)$ correspond to $\theta_1 = \sin^{-1}(n_{clad}/n_{core})$ and $\theta_2 = \pi/2$, respectively. The sensitivity of the sensor implemented with tapered fiber probe depends on the parameters such as taper ratio, taper profile, taper period, and taper length. Taper ratio is calculated by taking the ratio of the fiber core diameter (ρ) to tapered core diameter (ρ_0). By choosing a suitable taper ratio of the fiber, the sensitivity can be increased to a higher value. For all the taper profiles, as the taper ratio increases sensitivity increases. However, maximum sensitivity is obtained for the exponential-linear profile for a higher-taper ratio and minimum for parabolic profile.

The implementation of multitapered fiber-optic SPR probe for sensing is also reported (Srivastava and Gupta 2011). The sensitivity of the multitapered probe can be improved by choosing the optimized factors such as length of the total fiber probe, the length of sensing region with multiple tapering, taper period, taper ratio, linear regime length, and the number of tapers.

3.5.4 D-Shaped Probe

The D-shaped sensing region has a better cross-sectional shape which allows better interaction of the evanescent field with the sensing medium to improve the sensitivity of the sensor. A schematic of D-shaped fiber-optic probe is shown in Figure 3.9. In the case of D-shaped probe, the V-number is different from that of the usual straight probe and the entire D-shaped region plays the role of core (Gupta 2006, Chen et al. 2010, Qazi et al. 2016). The number of modes supported by the fiber in the D-regime is obtained using the operational V-number (V_{op}) (De-Jun et al. 2014) given by

$$V_{op} = \frac{2\pi\rho_{eff}}{\lambda}\left(n_{core}^2 - n_s^2\right)^{1/2} \tag{3.46}$$

where ρ_{eff} is the effective radius of the core in the D-shaped region and n_s is the refractive index of the sensing medium around D-shaped region. As the V-number of the sensing region changes, the rearrangement of power between the core and the sensing medium occurs. The depth/D-groove also affects the number of modes supported by the regime. An optimized groove/depth

FIGURE 3.9
Schematic of a D-shaped optical fiber probe.

of 500 μm in a fiber having core diameter of 1000 μm was reported in 2014 (De-Jun et al. 2014). The bending of the fiber is further possible in the D-shaped probe. The increase in bending helps in an enhanced distribution of power from core to sensing medium and hence an increase in the evanescent field absorption by the sensing medium (De-Jun et al. 2014). Thus, the factors affecting the sensitivity of a D-shaped probe are its groove/depth, length of sensing region, the number of D-grooves in the sensing region, and bending radius of the fiber.

3.6 Summary

In this chapter, we have presented an overview of the possibilities of implementing optical fiber as substrate for applications in environmental monitoring and biomedical applications to incorporate its real-time monitoring capability in the sensor-devising area. The fiber parameters that contribute to an efficiency hike in the performance of an optical-fiber-based sensor are discussed which helps one in easy analysis and choice of the parameters to design the best sensor. The performance of the sensor designed with all the optimized features is, generally, evaluated on the basis of various grading parameters which need to be better with easy realization processes in comparison to the competing sensors. The possibilities of transfiguring a straight uniform fiber core to tapered, U-shaped, and D-shaped structures enhance the sensitivity of the sensor.

References

Agrawal, H., A. M. Shrivastav, and B. D. Gupta. 2016. Surface plasmon resonance based optical fiber sensor for atrazine detection using molecular imprinting technique. *Sens. Actuators B* 227: 204–211.

Boef, G. D. and A. Hulanicki. 1983. Recommendations for the usage of selective, selectivity and related terms in analytical chemistry. *Pure Appl. Chem.* 55: 553–556.

Bogue, R. 2011. Fiber optic sensors: A review of today's application. *Sensor Rev.* 31: 304–309.
Bosch, M. E., A. J. R. Sanchez, F. S. Rojas, and C. B. Ojeda. 2007. Recent development in optical fiber biosensors. *Sensors* 7: 797–859.
Busurin, V. I., A. S. Semenov, and N. P Udalov. 1985. Optical and fiber-optic sensors (Review). *Sov. J. Quantum Electron.* 15: 595–621.
Cao, D., B. Shi, H. Zhu, G. Wei, S. E. Chen, and J. Yan. 2015. A distributed measurement method for in-situ soil moisture content by using carbon-fiber heated cable. *J. Rock Mech. Geotech. Eng.* 7: 700–707.
Chen, C. H., T. C. Tsao, J. L. Tang, and W. T. Wu. 2010. A multi-D-shaped optical fiber for refractive index sensing. *Sensors* 10: 4794–4804.
Culshaw, B. and A. Kersey. 2008. Fiber-optic sensing: A historic perspective. *J. Lightwave Technol.* 26: 1064–1078.
D'Amico, A. and C. D. Nitale. 2001. A contribution on some basic definitions of sensors properties. *IEEE Sens. J.* 3: 183–190.
Danzer, K. 2001. Selectivity and specificity in analytical chemistry. General consideration and attempt of a definition and quantification. *Fresen. J. Anal. Chem.* 369: 397–402.
De-Jun, F., Z. Mao-Sen, G. Liu, L. Xi-Lu, and J. Dong-Fang. 2014. D-shaped plastic optical fiber sensor for testing refractive index. *IEEE Sens. J.* 14: 1673–1676.
Ghatak, A. and K. Thyagarajan. 1998. *Introduction to Fiber Optics.* New York: Cambridge University Press.
Gupta, B. D. 2006. *Fiber Optic Sensors: Principles and Applications.* India: NIPA.
Gupta, B. D., H. Dodeja, and A. K. Tomar. 1996. Fiber-optic evanescent field absorption sensor based on a U-shaped probe. *Opt. Quant. Electron.* 28: 1629–1639.
Gupta, B. D., A. Sharma, and C. D. Singh. 1993. Evanescent wave absorption sensors based on uniform and tapered fibers: A comparative study of their sensitivities. *Int. J. Optoelectron.* 8: 409–418.
Gupta, B. D., A. M. Shrivastav, and S. P. Usha. 2016. Surface plasmon resonance-based fiber optic sensors utilizing molecular imprinting. *Sensors* 16: 1381.
Gupta, B. D. and C. D. Singh. 1994a. Evanescent-absorption coefficient for diffuse source illumination: Uniform- and tapered-fiber sensors. *App. Opt.* 33: 2737–2742.
Gupta, B. D. and C. D. Singh. 1994b. Fiber-optic evanescent field absorption sensor: A theoretical evaluation. *Fiber Int. Opt.* 13: 433–443.
Gupta, B. D., S. K. Srivastava, and R. Verma. 2015. *Fiber Optic Sensors Based on Plasmonics.* Singapore: World Scientific Publishing Company.
Hartung, A. and H. Bartelt. 2011. Light propagation in tapered optical fibers: Spatial light confinement and generation of plasmonic waves. *PIERS Proceedings,* Marrakesh, Morocco, March 20–23. 255–258.
Hecht, J. 1997. *Understanding Fiber Optics.* (2nd ed.). New Delhi: BPB Publications.
Iten, M., F. Ravet, M. Niklès et al. 2009. Soil-embedded fiber optic strain sensors for detection of differential soil displacements. *Proc. 4th Int. Conf. on Structural Health Monitoring of Intelligent Infrastructure (SHMII-4).* July 22–24 , 169.
Jorgenson, R. C. and S. S. Yee. 1993. A fiber-optic chemical sensor based on surface plasmon resonance. *Sens. Actuators B* 12: 213–220.
Kant, R., R. Tabassum, and B. D. Gupta. 2016. Fiber optic SPR-based uric acid biosensor using uricase entrapped polyacrylamide gel. *Photon. Technol. Lett.* 28: 2050–2053.

Kerttula, J., V. Filippov, V. Ustimchik, Y. Chamorovskiy, and O. G. Okhotnikov. 2012. Mode evolution in long tapered fibers with high tapering ratio. *Opt. Express* 20: 25461–25470.

Khijwania, S. K. and B. D. Gupta. 2000. Maximum achievable sensitivity of the fiber optic evanescent field absorption sensor based on the U-shaped probe. *Opt. Comm.* 175: 135–137.

Lagakos, N., J. H. Cole, and J. A. Bucaro. 1987. Microbend fiber-optic sensor. *Appl. Opt.* 26: 2171–2180.

Leung, A., P. M. Shankar, and R. Mutharasan. 2007. A review of fiber-optic biosensors. *Sens. Actuators B* 125: 688–703.

Long, G. L. and J. D. Winefordner. 1983. Limit of detection: A closer look at the IUPAC definition. *Anal. Chem.* 55: 712A–724A.

Qazi, H. H., A. B. Mohammad, H. Ahmad, M. Z. Zulkifli, and S. W. Harun. 2016. Single-mode D-shaped optical fiber sensor for the refractive index monitoring of liquid. *J. Mod. Opt.* 63: 750–755.

Ramane, D. V., S. S. Patil, and A. D. Shaligram. 2015. Detection of NPK nutrients of soil using fiber optic sensor. *Proc. IJRAT "ACGT 2015"*. E-ISSN: 2321–9637: 66–70.

Roriz, P., O. Frazao, A. B. Lobo-Ribeiro, J. L. Santos, and J. A. Simoes. 2013. Review of fiber-optic pressure sensors for biomedical and biomechanical applications. *J. Biomed. Opt.* 18: 050903.

Shrivastava, A. and V. Gupta. 2011. Methods for the determination of limit of detection and limit of quantification of the analytical methods. *Chron. Young Sci.* 2: 21–25.

Shrivastav, A. M. and B. D. Gupta. 2016. SPR and molecular imprinting based fiber optic melamine sensor with high sensitivity and low limit of detection. *IEEE J. Sel. Topics Quantum Electron.* 22: 6900207.

Singh, S. and B. D. Gupta. 2012. Fabrication and characterization of highly sensitive surface plasmon resonance based fiber optic ph sensor utilizing high index layer and smart hydrogel. *Sens. Actuators B* 173: 268–273.

Sony, K. and M. Soumya. 2013. Preparation of tapered optical fibers to utilize the evanescent field for sensing application. *Int. J. Eng. Trends Technol.* 4: 442–446.

Srivastava, S. K. and B. D. Gupta. 2011. A multitapered fiber-optic sensor with enhanced sensitivity. *IEEE Photon. Technol. Lett.* 23: 923–925.

Subrahmanyam, N., B. Lal, and M. N. Avadhanulu. 2012. *A Textbook of Optics*. (25th ed.). Delhi: S. Chand and Company Ltd.

Takeo, T. and H. Hattori. 1982. Optical fiber sensor for measuring refractive index. *Jpn. J. Appl. Phys.* 21: 1509–1512.

Takeo, T. and H. Hattori. 1992. Silica glass fiber photorefractometer. *Appl. Opt.* 31: 44–50.

Verma, R. K., A. K. Sharma, and B. D. Gupta. 2008. Surface plasmon resonance based tapered fiber optic sensor with different taper profiles. *Opt. Commun.* 281: 1486–1491.

Vessman, J. 1996. Selectivity or specificity? Validation of analytical methods from the perspective of an analytical chemist in the pharmaceutical industry. *J. Pharm. Biomed. Anal.* 14: 867–869.

Wang, L., A. Tomlin, S. Pandita et al. 2016. In-situ monitoring of cross-linking reactions using E-glass fibers and evanescent wave spectroscopy. *Sens. Actuators B* 236: 358–366.

4

Nanostructured Sensors

The term "nano," derived from the Greek word nânos meaning dwarf, is generally accepted for materials having dimensions within 100 nm. The distinctive materialistic, catalytic, and organic properties of the structures in nanoscience supported inventing new technology, which combines multiple field of study in science and engineering. In 1959, Richard Feynman gave a lecture in a session of American Physical Society where he explicated "There is Plenty of Room at the Bottom," introducing his ingenious ideas of the nano-world. The trend of integrating more and multiple electronic components on to a single chip was put forward by Gordan Moore in 1965. The Intel cofounder also showed a graph deploying Moore's law, which described the number of components per integrated function versus time, demonstrating his visionary of semiconductor nanoscale technology. In 1974, N. Taniguchi introduced the word nanotechnology for the mechanisms and assembling of materials in nano-range (Hulla et al. 2015). After 1980, the field of nanotechnology started developing and exploring possibilities in its full-fledge with the discovery of quantum dots, fullerenes, carbon nanotubes (CNTs), and much more. The concept of molecular nanotechnology was discussed by Eric Drexler in 1986 in his book "Engines of Creation: The Coming Era of Nanotechnology" in 1986 where he put forward the idea of nanoscale assembler. The building of nanostructures by synthesizing/fabricating materials of nanosize is, generally, performed by two approaches: top-down and bottom-up. The top-down approach includes breaking down of bulk structures to nanosize, while the bottom-up approach shows the rising of nanostructures by amassing the structures such as atom and molecules (Kalantar-Zadeh and Fry 2007). Although the top-down approach leads to an extent production of nanomaterials, the approach suffers from design and pattern imperfections in the nanostructures. The bottom-up approach provides easy realization for defectless structures in few nanometer-scale range. The combined usage of these approaches and their advantages made the field of nanotechnology ascertainable. In the twenty-first century, the world of nanoscience has started flourishing leading to the developments of technology, ideas, and works to bring forth the novel discussions and implementations. The growth of nanotechnology paved the way for industrial developments and hence the economic growth of the countries as well. It has been seen that, in the countries which hold a substantial lead in engineering, the technique of nanoscience has established a well-secured and balanced financial and civic background.

4.1 Nanotechnology as a Sensing Platform

The role of nanotechnology is extending day by day in physical, chemical, and biological fields. The techniques implemented by nanoscience are playing a lead role in various applications in agricultural, environmental, medical, health management, industrial, safety, and armament sectors. Sensing/sensor is one of the areas in which nanoscience holds an enormous impact. The contemporary approach needed in the sensing applications to support its growing requirements is made possible in an easy, profitable, and contented way with the help of nanotechnology. The performance of the nanotechnology-employed sensors has been found to be enhanced by multiple times as compared to the typical sensors in terms of sensitivity, specificity, cost-effectiveness, stability, regeneration, fabrication, and power/energy consumption. The peerless properties of materials fabricated using nanotechnology and the ability to modify the material properties specific to the oriented application make the technique remarkable among the competing techniques. Despite the transducer system, nanotechnology can be implemented to embellish the sensor integrity, which provides the sensor a better LOD and a simplified mechanism with label-free detection. Although nanotechnology is being used for enabling mechanical, electrical, electrochemical, and optical sensors, the most commonly employed environmental and biological sensors adopt optical transducer system. The extent of the nanotechnological application is dependent on the choice of material, structure, mechanism, and manufacturing method. When materials transform from bulk nature to micro- and nanosized particles and structures, the optical, electrical, physical, chemical, and mechanical properties of the material change due to quantum confinement and chemistry of the interfacial reactions and these properties can be tuned by controlling the size and shape of the nanoparticle. The nanomaterials can be either naturally occurring or synthetically prepared with dimensions of zero as in quantum dots and nanoparticles. According to shapes, these have been given the names as nanotubes, nanowires, nanoprisms, nanosheets, nanodisks, nanoflowers, and dendrite structures. The organic molecules such as carbohydrates, lipids, amino acids, proteins, cellulose, DNA, and RNA come under the former category of natural nanomaterials, while the handmade polymers such as polyaniline and polythiophenes, CNTs, metal/metal oxide nanoparticles, and quantum dots come in the latter category. The 0D nanomaterials have all the dimensions within nano-range, which is <100 nm. For 1D nanomaterials, one of the three dimensions is not in nanoscale, while the other two are within 100 nm. Similarly, in 2D and 3D, the two and none of the dimensions are not in nanoscale. The idea of 0D, 1D, 2D, and 3D nanomaterials can be understood from Figure 4.1.

The nanomaterials exhibit quantum/electron confinement effects which are useful for sensing applications (Arregui 2009). In 0D nanomaterial, in

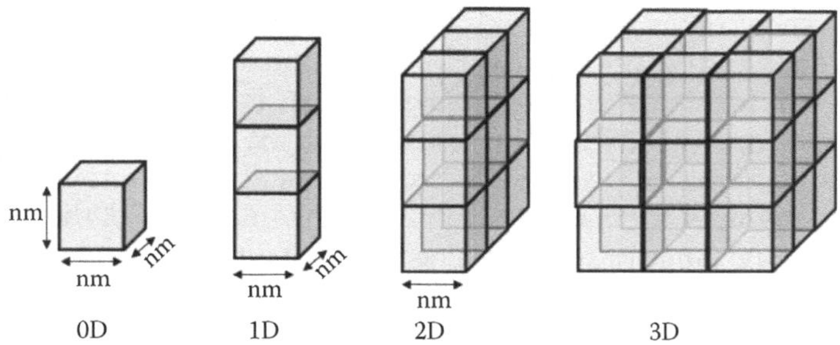

FIGURE 4.1
Schematic of 0D, 1D, 2D, and 3D nanomaterial demonstration.

which all the three dimensions are within the nanoscale, the electrons are fully constricted in the 3D expanse. In 1D materials, the electron delocalization occurs along the nonnano-dimensional axis, whereas electron confinement occurs in the other two. A complete electron delocalization occurs in 3D nanostructures. A detailed discussion on the synthesis and applications of these nanomaterials/structures is given in further sections.

Nanomaterials got eminent applications in environmental monitoring and biomedicine (Kalantar-Zadeh and Fry 2007, Filipponi and Sutherland 2012). The biomedical applications of nanomaterials include sensing, cancer detection, drug delivery, removal of toxic/unwanted foreign bodies, artificial bodily parts, implants, imaging, and tissue engineering (Malsch 2002). The nanomaterials that are commonly in use are that of metals, metal oxides, 2D carbon structures, transition metal dichalcogenides, silicate clays, layered double hydroxides, etc. The biocompatibility/nontoxicity and cytocompatibility of the nanoparticles make them support biomedical applications compared to the bulk structure. The operation of a nanomaterial depends on its shape, surface area, size, charge density, porous structure, surface chemistry, adaptability, and flexibility. The size of the nanostructure controls its movement inside the body in the case of *in vivo* application. For instance, CNTs and magnetic nanoparticles are used for drug delivery applications because of their easy transportation inside the blood circulatory system. The chemical reactiveness of the material depends on the surface area, surface-area-to-volume ratio, and the properties of the surface. The material surface regulates the interaction and the interface between the nanomaterial and environment determines the reactivity, catalytic activity, absorption, and adsorption. The surface area is enlarged in nanomaterials from that of a bulk material, keeping the total volume same. Thus, the aspect ratio, which is the ratio of the surface area to volume, increases. The energy of the electrons, atoms, and molecules is higher at the surface than at the core of the material, which again increases the reactivity. The higher energy makes the nanomaterial unstable innately, which is automatically tried to reduce by the nanoparticle itself by opting agglomeration (Filipponi and Sutherland

2012). The shape of the nanoparticle also affects the movement of nanostructure in blood vessels; due to the difference in the hydrodynamic forces, the different nanostructure endures. Materials can be microporous, mesoporous, or macroporous, depending on whether the diameter of the pore is <2, 2–50, or >50 nm, respectively. Microporous and macroporous structures are commonly used for biomedical applications since most of the medicinal agents are in the range of 2–50 nm. The flexible nature of nanomaterial eases its immobilization and passage through the circulatory system. However, nanoparticles also have some disadvantages. The chances for the nanoparticles to accumulate in the human body are more because of the slow degradation mechanism. This may affect the body parts such as lungs, kidney, etc. The reaction of these nanomaterials inside body may result in the production of free radicals that may cause oxidative damage as well (Chimene et al. 2015). The chemical nanospecies especially metal oxides generate reactive oxygen and the body tries to balance the condition, preventing the damages to cells, with the help of antibodies. The condition of imbalance created, if the body is not able to manage the oxidative damages, occurs due to the free radicals generated and is called oxidative stress. A higher level of oxidative stress results in cellular destruction and even in tissue damage. Hence, the choice of nanomaterials is crucial for both biomedical and environmental applications. The toxicity can be minimized by carefully selecting the material in its pure form with no contamination.

The principal environmental monitoring applications of nanomaterials in sensing include water, soil, food, and air quality control. Nanomaterials are used for the detection of food adulteration, pesticides, pollutants, pathogens, heavy metal ions in water, nutrients and pesticide level in soil, identifying the leakage of toxic gases, and other chemicals in industries. Detection is performed by absorption, adsorption, photocatalysis, oxidation–reduction, chemical binding, antibacterial activities, or creating by-products/radicals that change the sensing layer properties, which is identified and analyzed by the detector and analyzer of the designed sensor. This property of nanomaterials is also used for detoxifying the food compounds, water purification, wastewater treatment, and atmospheric remediation (Haes and Van Duyne 2003).

Water decontamination is a promising application of nanomaterials employing the ultrafiltration/nanoparticle-enhanced filtration activities such as nanomembranes, nanostructures with bioactivity, nanocatalysis, nano-sorbents, etc. In addition to this, the antibacterial properties of nanomaterials such as silver nanoparticles, dendrimers, ZnO, SnO_2 nanostructures, etc. are used in both biomedical and environmental applications including medicine, textile fabrics, water purification, food adulteration, implants surgery, sterilizing surgical devices/tools, and so on (Jan et al. 2014). The magnificent reductive properties of some nanomaterials are used to convert the highly noxious substances to less toxic ones, without the production of intermediate radicals/by-products which have after effects, and to decontaminate the solutions from toxic inorganic and organic contaminations. Among the factors that decide the application and implementation of nanomaterials, the synthesis and manufacturing methods are significant. The synthesis procedure of the nanostructure decides its physical and chemical properties such as shape, structure, solubility, conductivity, stability, and many more. Metal nanostructures are most commonly used for the sensor applications as they possess LSPR properties and high-catalytic activities. In the next section, we shall discuss about various methodologies for the fabrication of different metallic nanostructures.

4.2 Metallic Nanostructures and Synthesis

Over a century, the synthesis of metallic nanostructures and their applications in the industry have been an area of interest for the scientific community. These have been broadly used in variety of applications such as military, solar cells, electronic storage, drug delivery, biomedical diagnostics, plasmonic imaging, sensors, etc. The metallic nanostructures are synthesized and used in the abovementioned applications either directly or with the help of functionalization via various chemical routes which provide the ability to conjugate with the antibodies, ligands, drugs, and other functional groups of interest according to the application. Among metallic nanostructures, noble metallic nanostructures are most commonly used due to their versatile properties like inert nature, compatibility to biomaterials, and high stability. There are mainly eight noble metals. The most commonly referred noble metals are gold (Au), silver (Ag), platinum (Pt), and palladium (Pd). The noble metals are generally referred to as the atoms having the filled d-band of electronic structure. The properties of the bulk noble metal are markedly different from its nanostructure in terms of stability and reactivity. In chemical point of view, noble metals are inert to corrosion and oxidation in moisture which provide them stability. Due to the biocompatible, stable, and plasmonic properties, noble metal nanoparticles are extensively used for sensing applications as the transducer (LSPR, surface-enhanced Raman spectroscopy (SERS)), recognition agent, catalyst, etc. Therefore, in this section, we shall discuss various physical, chemical, and biological methods for the synthesis of different types of metallic nanostructures varying from 0D nanostructure like nanoparticle to 3D nanostructure like nanoflower. There are mainly two approaches for the synthesis of nanostructures: top-down and bottom-up. Figure 4.2 shows the difference in top-down and bottom-up approaches.

Top-down method refers to the cutting or slicing of the bulk sizes material up to the nanoscale size, while bottom-up corresponds to the building of the material from the atomic size toward the molecular, cluster, nanoscale-sized particle. Both the methods have their own pros and cons. The biggest

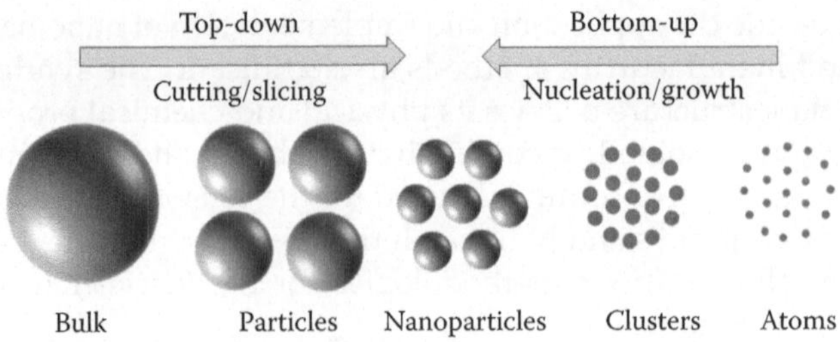

FIGURE 4.2
Concepts of top-down and bottom-up approaches for nanoparticles synthesis.

problems with top-down method are the presence of various imperfections in product structure and significant damage to the crystallographic planes, which lead to the various challenges for the device designing and its fabrication using nanostructures. Therefore, bottom-up approach is usually referred in nanotechnology. The growth of nanoparticles through bottom-up approach is called as the nucleation process. In the nucleation process, initially, the nucleus of the product material forms which acts as the seed material to produce the atoms to nanostructures. This may be referred to as the extreme localization of thermodynamic phases of nuclei which are the initial stages for the crystallization of materials. After that, the atoms start preparing the building blocks of atoms to atoms then molecules to molecules per their crystalline structure. LaMer was the one who explained the mechanism behind the nucleation process, known as LaMer mechanism (LaMer 1952). He synthesized the sulfur sols via the reduction of sodium thiosulfate. As per the LaMer mechanism, its process can be explained in three steps: (a) an increase in sulfur concentrations in the solution, (b) sulfur goes through the "burst nucleation process" to form the sulfur sols and the period of this process is the time up to which no further nucleation occurs, and (c) the diffusion process in which sulfur sols diffuse in the solution. Burst nucleation corresponds to the formation of a large number of product nuclei within a very short period, which is followed by the growth without any additional nucleation process. Thus, the product nuclei have similar growth parameters which allow the synthesis of product with same sizes. This is also referred to as homogenous nucleation. Any other case, if the nucleation and growth process are happening arbitrarily then the growth parameters are little different for all the nuclei so the heterogeneous size distributions would be obtained (Polte 2015). The shape, size, and various other properties of the product can be obtained from various characterization processes such as XRD, SEM, TEM, HRTEM, etc. Table 4.1 summarizes various characterization techniques used for confirming the morphology and hence the physical and chemical properties of the product.

Now we shall discuss various synthesis procedures of silver (Ag), gold (Au), platinum (Pt), and palladium (Pd) nanostructures.

TABLE 4.1
Various Techniques Required for the Characterization of Nanostructures

Characterization Technique	Basic Information Observation	Resolution	Remarks
X-ray diffraction (XRD)	Elements identification		Information about particle size, stress, strain can also be observed.
Electron-dispersive x-ray spectroscopy (EDX)	Elements identification		Simple and low cost.
Scanning electron microscopy (SEM)	Particle size and morphology	>20 nm	Low cost, easy to operate, sample preparation is also simple.
Field emission scanning electron microscopy (FESEM)	Particle size and morphology	>10 nm	No need for conducting coating as SEM, more resolved, element combination can also be found.
Transmission electron microscopy (TEM)	Particle size and morphology	>1 nm	Stress can also be observed. Dark-field imaging provides the reciprocal image hence the crystal nature.
High-resolution transmission electron microscopy (HRTEM)	Particle size and morphology	>1 nm	Stress, grain boundaries, dislocations, lattice vibrations can also be seen from micrograph.
Atomic force microscopy (AFM)	Particle size and topography	1 nm to 8 μm	Faster and cheaper than SEM and TEM.
Ultraviolet-visible (UV-vis) spectroscopy	Element identification, size, and shape		Concentration, agglomeration state, and refractive index near the nanoparticle surface can be found. Easy to operate.

4.2.1 Silver

Silver nanostructures have attracted researchers in the sensing applications due to their numerous distinctive physical and chemical properties along with high-thermal and -electrical conductivity, chemical stability, and non-linear optical behavior. Besides these, Ag nanostructures support plasmonic behavior, as discussed in previous chapters and act as transducer in LSPR- and SERS-based optical detection methods. In this section, we shall discuss various methods to synthesize Ag nanostructures. Every method has certain pros and cons based on the size, cost, stability, reactivity, homogenous size distribution, etc. First, we shall brief about the various chemical routes for the synthesis of Ag nanostructures.

4.2.1.1 Chemical Methods

The chemical route is the most extensively used for the synthesis of metallic nanostructures due to easy preparation process. For the synthesis of metal nanoparticles, generally, three main components are required: metal precursor, reducing element, and stabilizer. In the case of silver nanoparticle fabrication, usually, silver nitrate ($AgNO_3$) is used as the precursor which is reduced to Ag in the presence of specific reducing element. The reduction process has two steps: nucleation and subsequent growth. These steps control the size and shape of the synthesized Ag nanoparticle. For the preparation of nanoparticles with homogeneous size, the nucleation process of all the particles should start at the same time with same subsequent growth. These steps can be monitored by adjusting various reaction parameters such as selection of reducing agent and stabilizer, pH, reaction temperature, and pressure, etc. Polyol method is the most commonly used method for the synthesis of Ag nanoparticles.

Polyol method: In this method, monodisperse silver nanocubes having side width of about 50–115 nm range are prepared (Sun and Xia 2002a) via reduction of $AgNO_3$ with ethylene glycol at 160°C where polyvinylpyrrolidone (PVP) is used as the catalyst/stabilizer. The morphology of the prepared nanoparticles is greatly controlled by atmospheric conditions like reaction temperature, molar ratio of the units of PVP: $AgNO_3$ and $AgNO_3$ concentration. At temperatures >160°C, the produced nanoparticles take irregular shape. Further, for high concentration of $AgNO_3$, the resultant product takes the shape of nanocubes while for low concentrations (<0.1 M), silver nanowires are the main product. The shape of synthesized nanoparticles changes from nanocubes to truncated/untruncated nanoplates, spherical, circular nanodisks, depending on the molar ratio of PVP (Zou et al. 2007).

Wet chemical method: Wet chemical method has been used for the fabrication of silver nanorods over a glass substrate. In this method, initially seeds of Ag nanoparticles with sizes of 4 ± 2 nm are prepared in a solution. The seeds are prepared using $AgNO_3$ as precursor, $NaBH_4$ as reducing agent, and sodium citrate as the stabilizer. The prepared seed solution is then injected into an already prepared growth solution having silver nitrate, ascorbic acid, and cetyltrimethylammonium bromide (CTAB). The color of the solution changes to green after the whole reaction mechanism (Aslan et al. 2005). A little modification gives a colloidal solution of Ag nanostructure having spherical and rod shape, which can be separated by varying rpm and period during centrifugation (Orendoff et al. 2006).

Chemical vapor deposition: The synthesis of silver nanorods can be carried out using potassium tartaric for the reduction of silver nitrate in the presence of poly (vinylpyrrolidone) (PVP) (Gu et al. 2006). The diameter and the length of the silver nanorods can easily be controlled by varying $AgNO_3$/PVP ratio. Ag nanorods can also be synthesized using a simple chemical deposition route and porous aluminum membrane (PAM) (Xu et al. 2010). The nanorods are synthesized via the reduction of Ag^+ ions in PAM nanochannels by acetaldehyde. The concentration of PAM templates controls the diameter and length of the synthesized nanorods, while the reaction temperature controls the length of the nanorods.

Microemulsion method: Nanoparticles with more uniform particle size distribution can be prepared by microemulsion method. In this method, particles are prepared by mixing two organic solutions with different phases. In these two phases, one corresponds to the metal precursor and the other for reducing agent. The activity of interphase transportation between these two phases at their interface is mediated by a quaternary alkylammonium salt. The metal clusters prepared at the interface are very stable because of the coating of metal surface with the stabilizer molecule (Krutyakov et al. 2008). The use of highly deleterious organic solvents in this method is the most important drawback, which can be overcome by removing the surfactant and organic solvents from the final product.

Electrochemical method: In electrochemical method for the synthesis of silver nanoparticles, two electrodes are used. One is the silver anode and the other is the substrate cathode on which the nanoparticles are prepared. The electrodes are immersed into an electrolyte solution and the voltage is applied to both the electrodes. The silver is oxidized to silver ions as current flows to the silver anode. Silver ions travel to the substrate cathode and deposit on it as the nanoparticle due to the reduction process. The bunch of nanoparticles can then be separated by the vigorous stirring of the solution. The size and shape of the nanoparticles can be tailored by varying various parameters such as voltage, type of electrolyte solution, reaction period, etc. (Khaydarov et al. 2009). The composition of electrolyte solution alters the homogeneity of the nanoparticles in the solution.

Chemical synthesis of other nanostructure: Silver nanowires can be synthesized using the reduction of $AgNO_3$ through ethylene glycol at the reaction temperature of 160°C and adding a solution of $AgNO_3$/PVP. The finally added solution transforms the synthesized silver nanoparticles to the silver nanowires with 30–40 nm diameters and about 50 μm lengths (Liu et al. 2001). In the absence or low concentration of PVP, the resultant products are mainly nanospheres. The synthesis of silver nanowires in large quantity can be carried out by reducing the silver nitrate by trisodium citrate ($Na_3C_6H_5O_7$) in the presence of sodium dodecyl sulfonate (SDSN) using the seedless surfactant process (Hu et al. 2004). SDSN is used as the capping agent which leads to the formation of silver nanowires and only for a specific concentration of SDSN, silver nanowires are prepared. Silver nanowires with pentagonal shape are synthesized using polyol chemical method, where ethylene glycol (EG), PVP, and Fe (II) acetylacetonate are used as solvent/reducing agent, stabilizer, and catalyst (Wiley et al. 2006). Apart from nanospheres, nanocubes, nanorods, and nanowires, various complex silver nanostructures like nanoflower and nanopyramid have been synthesized using chemical routes. Pyramidical-shaped silver nanoparticles can be synthesized using polyol method where ethylene glycol (EG) is used as the solvent and sodium bromide (NaBr) as reducing agent. Although the chemical methods allow the synthesis of multiple simple nanostructures, the aggregation, uncontrolled particle size distribution, and the use of hazardous chemicals limit the applications of synthesized silver nanostructures.

4.2.1.2 Physical Methods

Physical methods for the synthesis of metal nanostructures are used when the particles with narrower size distribution and very less contamination are required. These are the two main advantages over the chemical synthesis process. The less contamination is because of the absence of solvent used in the nanostructure synthesis. In physical methods, the preparation of nanostructures and their depositions are generally performed using evaporation–condensation, laser ablation (pulsed laser deposition [PLD]), arc discharge method, evaporation, sputtering, and ion plating methods. A brief idea of these methods and their effects on the nanostructures synthesized are discussed in this section.

Evaporation–condensation: In evaporation–condensation method, the source material is used to keep in a tube furnace and heated at a specific temperature and atmospheric pressure so that the source material could vaporize and convert into its gaseous phase. When the vapor phases come out from the furnace through a carrier gas, it gets condensed to the solid phase as nanoparticles, nanocrystals, etc. Figure 4.3 shows the schematic of the method. The major disadvantage of the method is the requirement of large space for tube furnace (Kruis et al. 2000). This can be overcome by using a small ceramic heater for the evaporation of source material.

Laser ablation: Laser ablation is the process to remove the material nanoparticles from the source material by means of thermal treatment using a focused laser beam. In the process, the laser pulse with high intensity is irradiated over the source which provides the vapors of material to obtain the thermal equilibrium. The vapors are either condensed in an effective liquid medium (solvent) or are deposited over a substrate. The condensed product is normally nanostructure of the source material, while the deposited product is basically the thin film of the source material with few nanometer thicknesses. This process of film deposition is also called as PLD. The characteristics such as shape and size of produced nanostructures are strongly

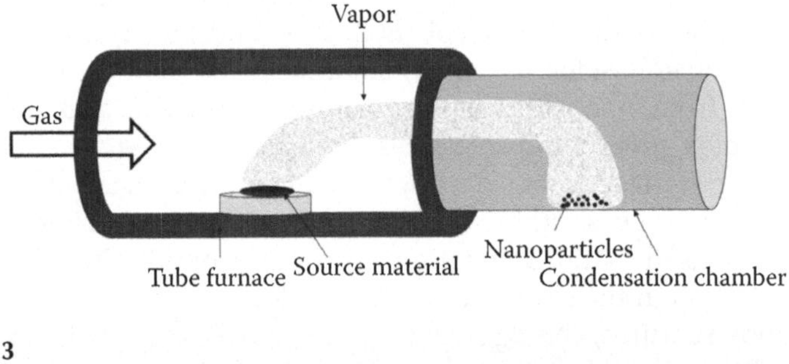

FIGURE 4.3
Schematic of evaporation condensation procedure depicting the nanoparticle synthesis by physical method.

influenced by many parameters such as wavelength and period of the impinging laser light on the source, beam diameter, ablation time, diameter of the laser pulse, and the presence/absence of the surfactant. The technique is advantageous due to the synthesis of nanostructures in the absence of any chemical solution which provides pure and uncontaminated product (Tsuji et al. 2002). The schematic of the laser ablation method for the synthesis of nanostructures is shown in Figure 4.4.

Various types of silver nanostructures prepared using this method and the influence of above-discussed parameters have also been studied (Mafuné et al. 2000a,b). In a study, silver nanospheres were fabricated using a femtosecond laser pulse of 800 nm wavelength. The particles were prepared in water and were of size around 20–50 nm (Tsuji et al. 2003). It was also found that the particles prepared with femtosecond laser pulse were significantly lower in size as compared to the particles prepared using the nanosecond laser pulse.

Arc discharge method: Arc discharge method is also one of the mostly used physical methods for the preparation of silver nanostructures. In this method, two electrodes of same source material are prepared with a specific potential difference. The electrodes are connected to provide an arc plasma, the arc plasma then vaporizes the source material because of the increase in the temperature gradient of the surrounding environment. The vaporized material, which is in the gaseous phase, condenses in some solvent like

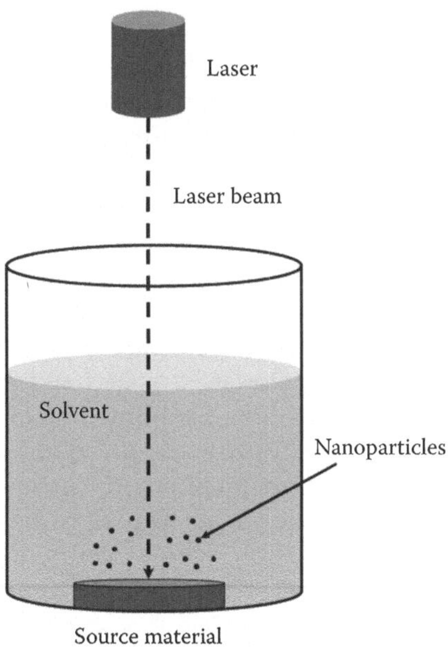

FIGURE 4.4
Process of laser ablation for nanostructure synthesis.

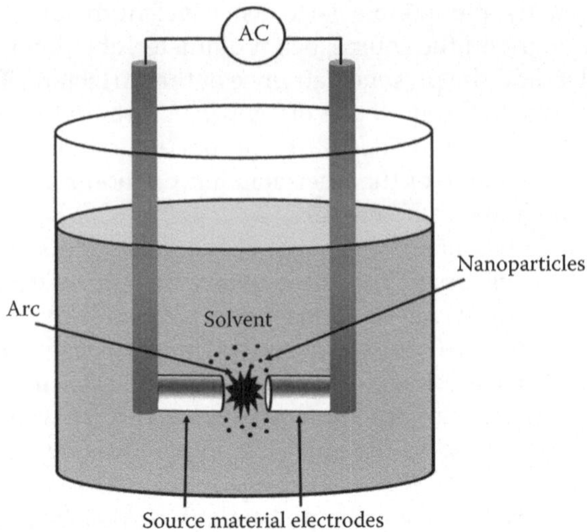

FIGURE 4.5
Schematic of arc discharge physical method for nanostructure synthesis.

deionized water. A pictorial representation of nanoparticle synthesis using arc discharge method is shown in Figure 4.5.

Evaporation: Evaporation is the most broadly used method for the preparation of the nanostructured film of various materials including metal and metal oxides. This is one of the oldest techniques used for the thin film coatings having the thickness up to the order of several nanometers. In this method, the source material is evaporated and the vapor is condensed over the substrate to give a coating of the thin film of the source material. Thin films with the large range of materials such as metals, metal oxides, and semiconductors, having different melting points, reactivity, and vapor pressure can be prepared. A variety of sources such as resistance heat filaments, electron beam, heating via conduction, and arcs are used for the evaporation of the target material. Thermal evaporation is the submethod among the numerous evaporation techniques and conventionally used for the deposition of nanofilms. The source material is melted and evaporated using an electric resistance heater. The vapors of sublimed material flow toward the substrate at the very low chamber pressure of around 10^{-6} mbar. A typical thermal evaporation coating system is shown in Figure 4.6.

The low chamber pressure allows the vapors to be deposited over the substrate without any contamination, which occurs due to the reaction or scattering with another molecule present in the chamber. The selection of the source material holder is such that it should not melt or sublime with the source material. Usually molybdenum (Mo: MP = 2623°C), tungsten (W: MP = 3422°C), and tantalum (Ta: MP = 3020°C) are used. The thickness and deposition rate of the film are monitored using a quartz crystal

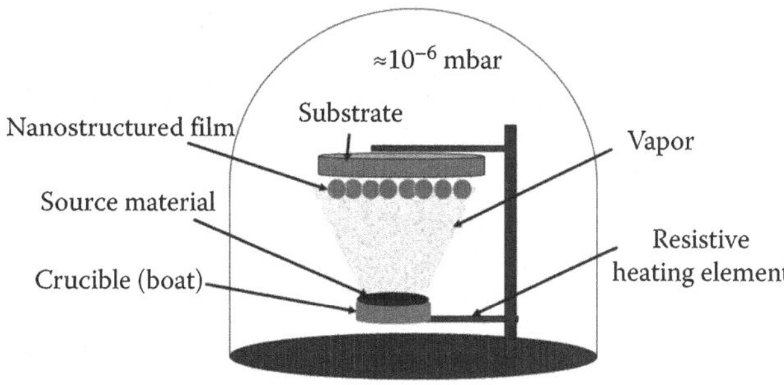

FIGURE 4.6
Pictorial representation of coating chamber of thermal evaporation system for nanoparticle synthesis.

microbalance (QCM) setup. The deposition rate is kept near 0.01 nm/s so that highly smooth film could be obtained (Usha et al. 2016). Another method for the fabrication of nanostructured thin films is electron beam evaporation. A high-energy electron beam from an electron gun source is used to fall over a small spot of the source shown in Figure 4.7.

In the process, a high voltage in the range of 1–100 kV is applied over the tungsten filament which results in the heating of the filament and production of electron beam. The beam is focused on to the crucible, which holds the source material and heats it up to evaporate. The evaporated material is then coated over the substrate. The path and the beam diameter are controlled via focusing and deflecting magnets. The cooling of the crucible is carried out by maintaining the continuous flow of water below its base (Campbell 2001). The whole process is performed under a very high vacuum of $<10^{-6}$ mbar. The chances of contaminations using this method are less than the thermal

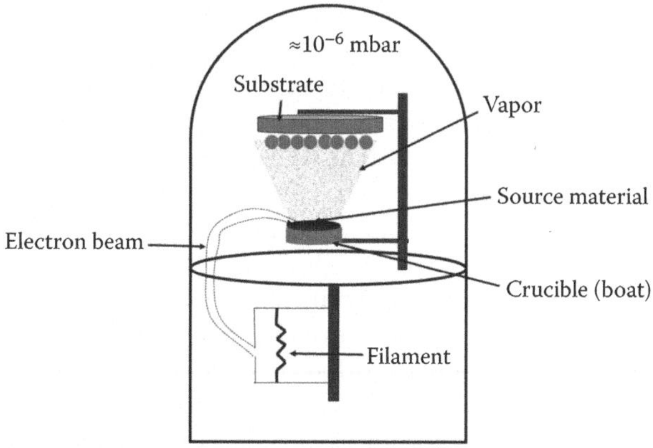

FIGURE 4.7
Electron beam evaporation setup for nanoparticle preparation.

evaporation method for the coating of the thin films. This is the most broadly used method for the thin film coating of various metals, dielectrics, and metal oxides. Few studies have been reported for the deposition of silver nanostructures using electron beam evaporation method, where Ag metal is used as the source material (Logeeswaran et al. 2008).

Sputtering: Sputtering is the process where the source material is bombarded with the high-energetic ions having approximate energy of 100 eV. Due to the bombardment, multiple collisions occur on the source material's surface, which result in the ejection of atoms of source material as the gas phase. These ejected atoms are directed toward the substrate over which its thin film is deposited. The number of atoms ejected from the source material surface per ion incident is called as the sputter yield. Various types of sputtering methods are used in the application for the thin film fabrication process such as DC sputtering, RF sputtering, magnetron sputtering, etc. Figure 4.8a shows schematic diagram of simple DC sputtering process, also called as diode sputtering.

In this process, the setup consists of cathode and anode under high vacuum. The cathode is made of the source material, while the substrate is kept over the anode. Further, a high voltage is applied between these electrodes in the presence of a suitable gas. Initially due to the high voltage, the gas breaks down to plasma and a dark electric field is created near the cathode which increases the number of collisions at the cathode. Thus, as discussed earlier, the atoms from the source surface sputter and move toward the anode and settle at the substrate. The atoms of source material should not react with the plasma of gas and hence usually the inert gas viz. argon (Ar) is used. If any oxide/nitride film is to be deposited, then the inert gas can be replaced by oxygen or nitrogen so that the chemical reaction between the source material and the oxygen/nitrogen gas can occur. This process is called reactive sputtering. Various parameters such as the distance between the source and substrate, softness of the source material, and applied voltage affect the deposition rate and uniformity of the coated film. In the case of RF sputtering, the concentration of gas plasma ions is controlled using the radio frequency power source (usually 13.56 MHz). The RF source produces more gas phase ions due to the alternate biasing between the cathode and anode at the higher rate (Ohring and Baker 2002). The properties of the plasma region can further be improved by using a magnetron source for the plasma generation. The method is known as magnetron sputtering, the setup of which is shown in Figure 4.8b. In this method, magnetic field is produced parallel to the cathode which forces the secondary electrons of the target source to trap near the cathode surface because the direction of electric field at the target is normal to the cathode and magnetic field is parallel to the surface. This results in the generation of highly dense plasma near the cathode surface, which improves the sputtering yield as well as the ionization of the sputtered gas. Therefore, the deposition rate gets enhanced. The magnetron setup also requires cooling

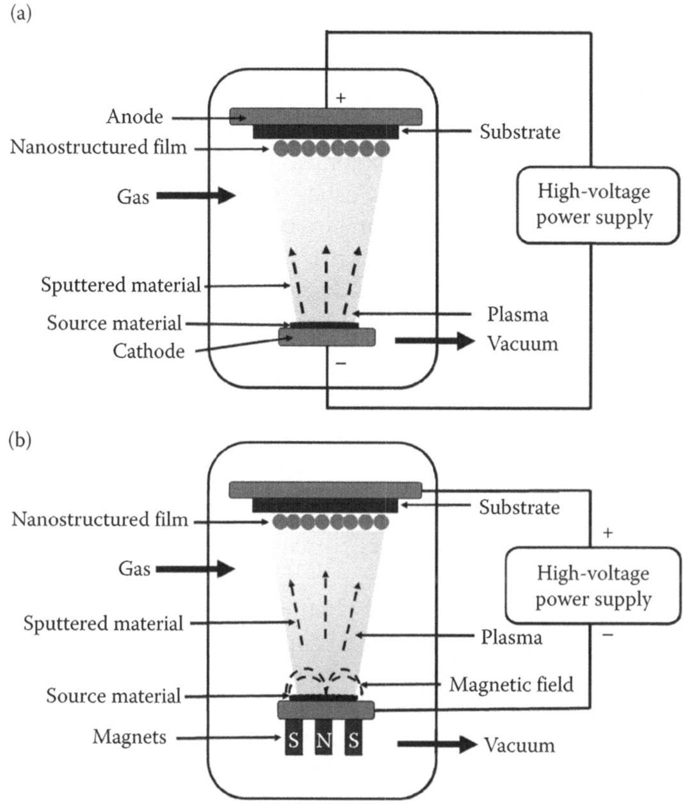

FIGURE 4.8
Schematic of (a) diode sputtering and (b) magnetron sputtering setup.

due to the production of thermal energy because of high collision of the electrons near the target. The properties of the nanostructured and nanometer-thick film can be controlled by various parameters of the sputtering process such as sputtering gas, sputtering power, distance between source and substrate, pressure, temperature, etc. Besides these, the substrate also plays an important role on the surface morphology and crystal structure of the product. Liquid sputtering is also a technique which is used for the synthesis of nanoparticles, where liquid substrate is used. The particle atoms travel from the cathode to liquid anode and in the liquid medium the vapor phase solidifies. Silver nanoparticles have been synthesized using liquid sputtering method (Siegel et al. 2012). The liquid medium in the study was glycerol (propane 1,2,3-triol). The method was quite fast as compared to the wet chemical method. The synthesized nanoparticles were having size of 3.5 ± 2.4 nm. The particles were round shaped and the size distribution of the nanoparticles was constant for diluted aqueous solutions up to glycerol-to-water ratio 1:20.

4.2.1.3 Biological Methods

The nanostructures synthesized via chemical methods are quite expensive and incompatible to the human body because of the use of toxic chemicals, while in the case of physical methods the instrumentation is far expensive. Thus, to get over these limitations, biological methods for nanostructure synthesis are introduced. These are also called as green synthesis of nanoparticles. The vast varieties of the organism are used for green synthesis, ranging from bacteria, fungi, algae, and plants (Mohanpuria et al. 2008). In the green synthesis protocol, the molecules produced from living organism work as the reducing agent as well as the stabilizer, while silver salts are used as the precursor. Below are briefly discussed various organisms.

Bacterium: A simplistic green synthesis method using the metal-reducing bacteria *Shewanella oneidensis* was proposed (Suresh et al. 2010). The bacterium was seeded with the $AgNO_3$ solution in a pH buffer. The synthesized nanospheres were within the size range of 2–11 nm and it was found that the pH of the buffer affects the size of the nanospheres. These particles had significant advantages such as biocompatibility, stability, hydrophilicity, and large surface area. Furthermore, the method is easy, reproducible, and requires less energy compared to the chemical methods.

Fungi: The green synthesis of silver nanoparticles was reported using fungus instead of bacteria. For example, the fungus *Trichoderma viride* showed the capability of the nanoparticle synthesis within the diameter ranging from 5 to 40 nm. Varieties of particle morphology were observed including spherical and rod-shaped structure (Fayaz et al. 2010). The enzymatic synthesis of silver nanoparticles using α-NADPH-dependent nitrate reductase obtained from *Fusarium oxysporum* and phytochelatin was reported (S. A. Kumar et al. 2007). In the study, silver nanoparticles were prepared with varying structure, size, and chemical compositions. Ag^{+2} ions were reduced in the presence of nitrate reductase resulting in the production of a stable silver hydrosol. The size of silver nanoparticles was around 10–25 nm, obtained with the help of capping peptides as a stabilizer. Further, the synthesis of silver nanoparticles was reported using a specific enzyme which makes their applications more advantageous. The main advantage of this technique is that it provides a new protocol to produce silver nanoparticles with a wide variety of particle morphologies without any aggregation.

Algae: A few protocols for the synthesis of silver nanoparticles using algal genera as the biological reagent have been reported. In these reports, marine algae such as *Isochrysis galbana, Sargassum boveanium,* and *Tetraselmis gracilis* were used (Shahverdi et al. 2007). The extracellular formation of silver nanoparticle with quite smaller size (7–16 nm) and spherical morphology by the exposure of Spirulina *platensis* biomass to aqueous $AgNO_3$ (10 M) solution at 37°C for 120 h was reported (Govindaraju et al. 2008).

Plant extract: In the biological synthesis of nanoparticles, the plants are the most commonly used biological reagent because these are very cost-effective for the production in large scale (Iravani 2011). In one of the studies reported for the synthesis of silver nanoparticles, green tea extract (*Camellia sinensis*) was used as the reducing agent and stabilizer (Vilchis-Nestor et al. 2008). Various types of phenolic acid compounds such as caffeine and theophylline present in *C. sinensis* were used for the production and stabilization of nanoparticles. Table 4.2 shows various methods used for the fabrication of silver nanostructures with different morphologies.

4.2.2 Gold

Apart from silver, gold is also one of the noble metals employed in sensing activities. Gold nanoparticles are the most stable among all the noble metal

TABLE 4.2

Various Methods for the Production of Ag Nanoparticles

Nanostructure Morphology	Size (nm)	Method	Reagents: Precursor Reducing Agent Stabilizer	Effecting Parameters	References
Nanospheres	2–11	Biological	$AgNO_3$ *Shewanella oneidensis* *Shewanella oneidensis*	pH of the solvent	Suresh et al. (2010)
Nanospheres	11	Electrochemical	$AgNO_3$ Ti cathode, Pt anode PVP		Lim et al. (2006)
Nanocubes	50–115	Chemical	$AgNO_3$ Ethylene glycol PVP	$AgNO_3$, PVP concentration, temperature	Sun and Xia (2002b)
Nanocolloids		Chemical	$AgNO_3$ Aqueous ammonia PVP	Molar ratio between Ag ion/PVP, temperature	Zou et al. (2007)
Nanorods	4 ± 2 (dia)	Chemical	$AgNO_3$ $NaBH_4$ Sodium citrate		Aslan et al. (2005)
Nanorods		Chemical vapor deposition	Ag^+ ions Acetaldehyde PAM nanochannels	Reaction temperature	Xu et al. (2010)
Nanowires	30–40 nm	Chemical	$AgNO_3$ Ascorbic acid		Zhang et al. (2004)

Notes: $AgNO_3$ = silver nitrate, $NaBH_4$ = sodium borohydride, PVP = polyvinylpyrrolidone, PAM = polyacrylamide.

nanoparticles. The high-chemical stability, biocompatibility, and the tuning of the chemical and optical properties with the particle size variation make its applications in a wide range of research areas such as electron transfer theory, biolabeling, catalysis, self-assembly, DNA melting, crystal growth, sensors, etc. In this section, we shall focus on the various physical, chemical, and biological strategies for the synthesis of gold nanostructures.

4.2.2.1 Chemical Methods

The chemical method involves three most important elements: precursor (salt), reducing agent, and the stabilizer. In the case of gold nanostructure synthesis, usually chloroauric acid ($HAuCl_4$) is used as the precursor, which is reduced to gold atoms and further growth of them leads to the synthesis of different nanostructures. Various types of reducing agents such as citrates and bromides can be used for these. Below, we shall discuss few of the chemical methods used for the synthesis of gold nanostructures.

Turkevich method: This is one of the most conventional methods for the gold nanoparticle synthesis and was suggested in 1951 (Turkevich et al. 1951). In this method, $HAuCl_4$ is reduced to gold nanoparticle in water using trisodium citrate, which acts as the reducing agent. Trisodium citrate also works as stabilizing agent in the reaction. The nanoparticles prepared using this method fall within the size range of 10–20 nm. The product particles are highly monodispersed and mostly spherical. A TEM micrograph of the gold nanoparticles prepared using this method is shown in Figure 4.9 (Srivastava et al. 2012).

The sequence of the change in the color of the resulting solution during preparation is usually pale yellow, colorless, very dark blue, purple, and finally the ruby red. Interestingly, the morphology of the colloidal solution at very dark blue color is the network of nanowires with 5 nm diameter, while the ruby red solution has nanoparticles of size range 10–20 nm (Pong et al. 2007). This might be because of the linear aggregation of the small gold nanocluster (5 nm diameter). As the aggregation increases, the nanowires are converted to nanoparticles with a higher diameter of 10–20 nm. The larger size nanoparticles can also be prepared using this method but with lesser amount of trisodium citrate (<0.005%) as it will not be sufficient to inhibit the aggregation of nanoparticles and thus the particles with higher size can be prepared. However, in such cases, the shape and dispersity of the product get altered and the prepared nanoparticles do not show monodisperse property (Kimling et al. 2006). The main disadvantage of Turkevich method is that the nanostructures synthesized are only electrostatically stable and in the presence of salts they cannot be retained for a longer time.

Brust method: Brust method is similar to the above method but instead of using sodium citrate as the reducing and stabilizing agent, sodium borohydride ($NaBH_4$) is used as the reducing agent and triacetylammonium bromide (TOAB) is used as the stabilizer. Toluene is used as the solvent which

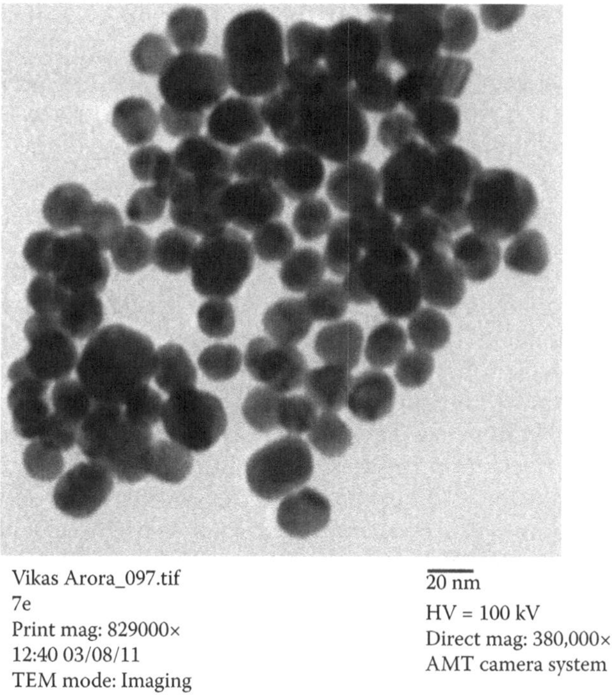

FIGURE 4.9
TEM micrograph of gold nanoparticles prepared by Turkevich method. (With kind permission from Springer Science+Business Media: *Plasmonics*, localized surface plasmon resonance based fiber optic U-shaped biosensor for the detection of blood glucose. 7: 2012, 261–268, Srivastava, S. K. et al.)

also protects gold nanoparticles from aggregation (Brust et al. 1994). The following reaction causes the synthesis of gold nanoparticles using this protocol:

$$3NaBH_4 + 4HAuCl_4 \rightarrow 4Au + 3NaCl + 3BCl_3 + 4HCl$$

The nanoparticles prepared using this method are in the range of 2–6 nm. It may be noted that TOAB is not a strong stabilizer and therefore the nanoparticle solution gets aggregated within 2 weeks after preparation. However, this could be overcome by using some strong stabilizer such as thiols (alkanethiols more specifically), but it affects the physical properties of the prepared gold nanoparticles (Manna et al. 2003).

Sonolysis: It is based on ultrasound. In this method, glucose is used as the reducing agent and reaction of $HAuCl_4$ with glucose occurs in the presence of ultrasonic waves (Zhang et al. 2006). The elements from the glucose hydroxyl radicals and sugar pyrolysis radicals work as the reducing agent. The prepared nanogold structure has a ribbon-shaped morphology with the width of 30–50 nm and few micrometer length. These ribbons have very flexible properties and can be bent >90°. It has been observed that while replacing the glucose by cyclodextrin, the product acquires a spherical morphology which signifies the role of glucose in the nanoribbon synthesis.

Seed-mediated synthesis: Gold nanorods with high-aspect ratio can be synthesized using the "seed-mediated method." The method involves the seed layers of synthesized nanoparticles (Jana et al. 2001). The solution of gold nanoparticles is prepared by reducing $HAuCl_4$ using a strong reducing agent such as $NaBH_4$. The prepared gold nanoparticles work as a seed layer which provides the nucleation sites for nanorod fabrication. Hence, by controlling the parameters of the seed solution, one can tailor the diameter as well as the aspect ratio of the final product of nanorod solution. In the nucleation process, the gold nanoparticles are extracted from the seed solution and kept in nanorods growth solution which contains $HAuCl_4$ and a weak reducing agent such as ascorbic acid (Cai et al. 2008). Besides seed-mediated method, various other methods such as electrochemical and photochemical have been reported for nanorod synthesis. Gold nanoshells with silica nanoparticles as the core can also be prepared using seed-mediated method with little modification in the abovementioned process. For this, if the seed layer of gold nanoparticles is changed by the silica nanoparticle, one can easily prepare the nanoshell structure of gold. The seeds of silica nanoparticles (core) can be fabricated by reducing the tetraethyl orthosilicate ($SiC_8H_{20}O_4$) in ethanol solution (Tadanaga et al. 2013).

4.2.2.2 Physical Methods

Following physical methods have been used for the fabrication of gold nanostructures:

Evaporation/Condensation method: Gold in a ceramic boat is kept in a tube furnace and is heated up to 1400°C, which causes the evaporation of gold in aerosol form. The gold aerosol is then carried using nitrogen, flowing in the tube furnace (flowing rate with 10 L/min), to the condensation chamber where the condensation of aerosol happens and the precipitation of gold nanoparticles with the maximum diameter of about 30 nm occurs (Magnusson et al. 1999).

Laser ablation method: The size-selected preparation of the gold nanoparticles using laser ablation method has been reported (Mafuné et al. 2002). The laser beam (552 nm) is irradiated over gold in SDS solvent. SDS works as a surfactant which prevents the particle aggregation. The particles prepared are small (1.7–5.5 nm), which can be tuned by the properties of the laser beam such as beam diameter, beam intensity, etc.

Arc discharge method: In this method, arc is produced by touching two electrodes of the same source material. The particles extracted through the arc are condensed with the solvent and lead to the preparation of nanoparticles (Lung et al. 2007). For the synthesis of gold nanoparticles, two gold rods (electrodes) with about 1 mm diameter are immersed in ethanol solution (solvent), which also works as a surfactant. By this method, the prepared nanoparticles have a size distribution of 2–40 nm (Tseng et al. 2009). Numerous other physical methods such as evaporation, sputtering, ion platting are also reported in the literature which show that the size and shape of the nanoparticles are strongly dependent on the parameters of the preparation method.

4.2.2.3 Biological Methods

Many biological strategies are reported in the literature for the formation of gold nanostructures using biomolecules, as the product is nontoxic, biocompatible and due to the ability for large-scale product generation. As mentioned earlier, numerous types of reducing agent such as bacteria, fungi, plants, etc. are used for the synthesis. Common bacteria used for gold nanoparticle synthesis are *Pseudomonas fluorescens* and *Geobacillus stearothermophilus*. In the case of fungi for gold nanoparticle fabrication, most commonly used fungus is *Candida albicans* and the nanoparticles synthesized have size in the range of 20–80 nm. Plants have various types of biocomponents such as quinones, phytosterols, flavonoids, etc., which can lead the synthesis of gold nanoparticles as they have the functional groups which can be used for increasing the reduction rate and for increasing the stabilization. Due to this, various researchers have proposed the synthesis of gold nanoparticles using extracts of various plants like *Sapindus mukorossi, Prunus domestica,* etc. Table 4.3 shows various methods reported in the literature for the synthesis of gold nanostructures.

4.2.3 Platinum

Platinum is the next metal lying in the category of noble metals which is costly due to its rare availability. The nanostructures of platinum are usually prepared in the form of colloids or suspension with black or brownish red color. Few synthesis protocols for Pt nanostructures are discussed below.

4.2.3.1 Chemical Methods

There are several chemical methods for the preparation of Pt nanostructures that have different morphologies. In these methods, usually Pt precursor is reduced to Pt atoms with the help of reducing agent and the prepared Pt atoms begin to aggregate in the form of nanosized particles.

TABLE 4.3

Physical, Chemical, and Biological Methods for Gold Nanostructure

Nanostructure Morphology	Size (nm)	Method	Reagents: Precursor Reducing Agent Stabilizer	Effecting Parameters	References
Nanospheres	1.7–5.5	Physical	Au metal Laser beam SDS	Beam diameter and power	Mafuné et al. (2002)
Nanosphere	2–40	Physical	Au metal electrode Arc Water (solvent)	Arc period	Tseng et al. (2009)
Nanowires	5 nm (dia)	Chemical	$HAuCl_4$ Trisodium citrate	Trisodium citrate concentration	Kimling et al. (2006)
Nanoislands	10–20 25–180	Physical	Au metal e-beam evaporation		Gaspar et al. (2013)
Nanoribbons	30–50 nm (width)	Chemical	$HAuCl_4$ Glucose	Glucose is responsible for ribbons	Zhang et al. (2006)
Nanostars	50	Chemical	$HAuCl_4$ DMF PVP	PVP concentration, reaction temperature	P. S. Kumar et al. (2007)

Notes: $HAuCl_4$ = chloroauric acid, $NaBH_4$ = sodium borohydride, SDS = sodium dodecyl sulfate, PVP = polyvinylpyrrolidone, DMF = dimethyl formaldehyde.

A capping agent or stabilizer is also used for the stabilization of the nanostructures and preventing them from further aggregation. Most commonly used precursors for Pt nanostructure synthesis are chloroplatinic acid (H_2PtCl_6) and platinous chloride (K_2PtCl_6). Few examples of the reducing agent include sodium borohydride ($NaBH_4$), hydrogen gas (H_2), and ethylene glycol ($C_2H_6O_2$), and generally used stabilizers are sodium citrate and polyacrylic acid (Islam et al. 2014). In the colloidal method, Pt nanoparticles can be fabricated through the reduction of H_2PtCl_6 (precursor) by alcohol (reducing agent) in the presence of polyacrylic acid (polymeric stabilizer/capping agent) (Humphrey et al. 2007). Usually, Pt nanoparticles with hexagonal shape (1–8 nm) are found as the product of this method. But the size and shape can be tuned by varying the molar concentration and growth parameters which may change the particles shape from hexagonal to cubic or some intermediate shapes named as cuboctahedra (similar to cubic but with truncated vertices). It has been observed that the particle size and morphology strongly depend upon the solvent, the reaction atmosphere, capping agents and their concentrations, precursor, and the atmospheric conditions. The other nanostructures such as nanowires and nanorods can be fabricated by using the synthesized nanoparticles as seed layer and with slow reduction of precursor solution in the presence of the seed particles.

4.2.3.2 Physical Methods

There are several physical methods reported for the synthesis of Pt nanoparticles such as sputtering, ion platting, and laser ablation. In the sputtering method, the uniformly distributed spherical Pt nanoparticles can be prepared over tilted substrate using low electrical power under high vacuum (Jeff et al. 2011). Using this method, ultrasmall nanoparticles ranging from 0.7 to 1.34 nm are fabricated. The laser ablation approach involves the irradiation of laser beam (532 nm, 1.2 mm diameter) over the Pt metal plate in water solution having SDS as the surfactant. As product, Pt nanoparticles with uniform size of about 6.2 nm are obtained. The average diameter of Pt nanospheres strongly depends upon the SDS concentration.

4.2.3.3 Biological Methods

Unlike gold and silver, a very less information is available on the green synthesis of Pt nanoparticles. A study using the fungus, *Neurospora crassa*, for the synthesis of Pt nanoparticles was reported with the product having the spherical morphology and 4–35 nm diameter (Castro-Longoria et al. 2012). The size of synthesized Pt nanoparticles depends on the reaction parameters such as temperature and extract concentration. Extract from green tea and black tea having large polyphenolic concentration may also be employed as the reducing/capping agent for the fabrication of Pt nanoparticles. Table 4.4

TABLE 4.4

Various Synthesis Protocols for Pt Nanostructures

Nanostructure Morphology	Size (nm)	Method	Reagents: Precursor Reducing Agent Stabilizer	Effecting Parameters	References
Nanosphere	0.7–1.34	Physical	Pt metal	Arc properties	Jeff et al. (2011)
Hexagonal and cubic	2–6	Chemical	H_2PtCl_4 Alcohol Polyacrylic acid	Molar concentration of H_2PtCl_4	Humphrey et al. (2007)
Nanosphere	2–12	Biological	H_2PtCl_4 *D. kaki plant* *D. kaki plant*	Reaction temperature	Song et al. (2010)
Nanosphere	4–35	Biological	H_2PtCl_4 *N. crassa* *N. crassa*		Castro-Longoria et al. (2012)

Note: H_2PtCl_4 = chloroplatinic acid.

shows numerous physical, chemical, and biological methods reported in the literature with varying morphology of the nanostructures.

4.2.4 Palladium

Palladium group of metals (PGM) is the group of metals having high-catalytic activity with their application as the catalyst for hydrogenation and dehalogenation processes. Palladium is the most important member of PGM family. Its physical and chemical properties change in the presence of hydrogen gas; thus, it is widely used as a sensing medium for hydrogen gas (Tabassum and Gupta 2015a).

4.2.4.1 Chemical Methods

Here, we shall discuss only one chemical method for the preparation of palladium (Pd) nanostructures.

Organic synthesis: In this method, usually potassium tetrachloropalladate (K_2PdCl_4) is used as the precursor, which is reduced in the presence of polyethylene glycol (PEG) that acts as a stabilizer and prevents the agglomeration of the nanostructures. The method provides the easy, cost-effective, and fast synthesis of Pd nanoparticles where the size can be tuned by various parameters such as Pd:PEG ratio, PEG concentration in water, etc. The size distribution of particles usually lies within the range of 5–11 nm (Sawoo et al. 2009). Apart from this, several other chemical approaches are reported in the literature for the synthesis of Pd nanoparticles using different types of the precursor, reducing agent, and stabilizer, which propose that the size and particle morphology can be easily varied by changing the atmospheric conditions and growth parameters.

4.2.4.2 Physical Methods

Laser ablation method: Colloidal solution of Pd nanoparticles is prepared by using the laser beam (1064 nm) irradiated over the Pd foil submerged in the water. The Pd atoms ablate from the foil and condensed within the solution. The particles formed range from few nanometers to 40 nm with round shape (Boutinguiza et al. 2016). Using other physical methods, such as arc discharge, evaporation–condensation, and sputtering, synthesis of nanocomposites of palladium can be performed. For example, employing the arc discharge method, CNTs filled with Pd nanoparticles can be fabricated (Bera et al. 2004). In this method, two electrodes of graphite (diameter: 3.05 and 12.70 mm) were immersed in 2.087 mM $PdCl_2$ solution. The electrodes are brought together and an arc is produced for a fraction of second. The solution is then cooled using water bath maintained at 7°C. The solution contains CNTs encapsulated with Pd nanoparticles with a particle diameter of 3 nm. Similarly, various complexes Ag–Pd and Au–Pd are reported in the literature using thermal evaporation and evaporation–condensation method, respectively.

4.2.4.3 Biological Methods

Like Pt, very few studies are known for the biological synthesis of Pd nanostructures. A green synthesis method for the preparation of Pd nanoparticles has been reported using Alpine sites of bacteria having the impurities of heavy metal ions. The bacterium with heavy metals contains *Pseudomonas* cells which work on reduction or on dichlorination of potassium tetrachloropalladate (K_2PdCl_4). After reduction, zero-valent-nonaggregated Pd nanoparticles can be found (Schlüter et al. 2014). Table 4.5 shows the methods used for palladium nanostructures by physical, chemical, and biological methods.

4.3 Effect of Nanostructures

In the optical field, sensors are the main area which inherits the advances happening in applications of nanotechnology in day-to-day life. In other way, these fields of applications demand the development of nanomaterials and nanostructures with an accurate regulation over their dimensions such as size, shape, alignment, and orientation. This is because the changes in these dimensions change the optical, electrical, catalytic, and physical properties of the structure which differ for different materials, compositions,

TABLE 4.5
Methods for Pd Nanostructure Fabrication Reported in Literature

Nanostructure Morphology	Size (nm)	Method	Reagents: Precursor Reducing Agent Stabilizer	Effecting Parameters	References
Nanosphere	1.4–6.8	Chemical	K_2PdCl_4 $NaBH_4$ Mercaptoammonium		Cliffel et al. (2000)
Nanosphere	3	Physical	Pd electrode	Cooling temperature	Bera et al. (2004)
Nanodisk	5–11	Chemical	K_2PdCl_4 Polyethylene glycol (PEG) PVP	Pd:PVP ratio	Sawoo et al. (2009)

Notes: K_2PdCl_4 = potassium tetrachloropallidate, $NaBH_4$ = sodium borohydride, $Pd(acac)_2$ = palladium acetylacetonate, TOP = trioctylphosphine.

and combinations. The sensing mechanism using nanomaterials is chosen as per the properties and capabilities of the materials and their structural properties. The sensing mechanism is because of the interaction of the nanomaterial surface with the analyte. The quantum confinement in 0D to 3D nanomaterials results in changes in the bandgap energy. An increase in the bandgap energy occurs during this quantum confinement which results in high demand of energy supply to the material. As the materials go down to their smaller dimensions, the quantization of energy levels occurs which in turn changes the optical properties of the nanomaterial such as absorption and emission that can be tuned in terms of the oriented application. Nanomaterials used in sensing applications are either conductors or semiconductors (e.g., CNT). The optical properties of the nanostructures, especially nanoparticles, depend on the size, shape, aspect ratio, and structures. For example, in the case of LSPR (discussed in Chapter 2), the resonance properties depend on the nanoparticle size. If the size of particle is less than λ, the wavelength of light, then the absorption exceeds the scattering in extinction coefficient and vice versa. For LSPR sensors, the change in refractive index of the sensing medium is considered as a calibration parameter which changes with the analyte properties. Here, the refractive index also depends on the composition and combination of the material, nanoparticle/nanostructure size, and the aspect ratio. The sensor sensitivity increases with an increase in the aspect ratio and by choosing certain metals/organic molecules to make complex compositions with nanostructures which enhance their properties. The shape and size of the nanomaterial/nanostructure influence the sensitivity of sensor. The nanostructures having sharp edges result in greater sensitivity to refractive index than blunt-shaped ones. As the complexity of the nanostructure increases, the intensity of plasmonic spectra and the sensitivity of the sensor increase. Other optical properties of nanomaterials which support sensing include fluorescence, photoluminescence, chemiluminescence, photoconduction, scattering, absorption, plasmonic resonance and lossy mode resonance, the phenomena that have already been explained in Chapters 1 and 2. Nanomaterials are also used as fluorophores in sensors. The plasmon and lossy mode resonance are used either as the main transduction mechanism or for enhancing the performance in terms of sensitivity of the sensor (Arregui 2009). The metallic nanoparticles of gold and silver are reported to exhibit the plasmonic activities and hence high-resonance characteristics in the visible region, which support the usage of these as single-nanoparticle-enabled nanosensors for biomedical and environmental monitoring applications (Sriram et al. 2015). The sensitivity of these resonances in turn depends on the nanomaterial properties as discussed earlier.

In nanoscale, materials are sensitive to even very small change in their mass. This leads to the implementation of nanostructures-based mechanical sensors as well. For example, the mechanical strength of gold nanowires is reported to be excellent enough to employ it for designing the pressure sensors, while the mechanical properties of ZnO nanowire such as piezoelectricity and noncentral symmetric structure support the displacement sensing and so on.

4.3.1 Shape of Nanoparticles

The shape and size of the nanoparticle decides the application and affects the performance/response of the sensor in precise. A change in the length of one of the dimensions of the structure occurs due to different reasons in different materials. In the case of metallic nanoparticles, a reduction in their size and shape causes an electron collective oscillation in the conduction band from surface to surface of the particle. This collective oscillation frequency results in an enhanced absorption of visible light. However, in the case of semiconductors, this leads to a confinement of the motion of electrons in a length that is comparable or smaller than that of the electron Bohr radius. A difference in the surface energy can also be seen in particles of different shape. Especially in the case of LSPR-based sensors and the surface-enhanced sensors using LSPR, the shape of the nanoparticle plays a considerable role. For instance, considering gold nanosphere and nanorod structure, the nanosphere exhibits a single resonance, while the nanorod or nanoparticle of rod shape exhibits two resonances due to the plasma oscillations along the long as well as the short axis which in turn affect the performance of the sensor utilizing them. Thus, the length-to-width ratio or surface-area-to-volume ratio, generally termed as aspect ratio, is an important factor that affects the potential and effect of nanostructures.

4.3.1.1 Spherical

In a spherical particle, the number of atoms on surface per unit volume is minimum and hence it is the stable structure among all the nanostructures in a thermodynamic way. The surface-area-to-volume ratio of the spherical nanoparticle is $3/R$, where R is the radius of the particle. At the same time, a smaller spherical particle has a higher surface energy and curvature which makes it highly unstable. Another feature of the spherical structure is the higher-ordered crystal planes on the surface. However, as the density of atoms on surface per unit volume increases, the instability of the structure increases, which happens with the structures such as tetrahedron, octahedron, and polyhedron.

The environmental factors such as substrate, attachment, or immobilization layers on the top and bottom of nanostructure, the target analyte solvent, the distance between particles, and so on affect the optical properties of the nanostructure. For example, a spherical particle of size >30 nm, the wavelength at which plasmon resonance (LSPR) occurs, increases due to the enhanced electromagnetic effects by almost 50 times in the direction of polarization. This makes the technique useful in SERS for electromagnetic field enhancement as well. Similarly, as the shape of nanoparticle changes, the aspect ratio changes and hence the plasmon resonance wavelength (Kelly et al. 2003). In addition to the basic properties such as absorption, scattering, and extinction, other characterizations such as LSPR, SERS, etc. depend on the structure of the nanomaterial. The change in the nanoparticle shape from spherical to rod, rod to disk, and disk to complicated ones increases the plasmonic bands from one to two, two to three, and three to more. It has been studied that for calorimetric detection of target analyte using LSPR technique, the nanoparticle used in the solution or matrix prepared should have mostly spherical nanoparticles (Stewart et al. 2008). Scattering properties also vary with the nanostructure. For instance, smaller nanoparticles of zinc oxide do not scatter visible light, while larger and elongated particles scatter.

4.3.1.2 Others

There are nanostructures other than simple nanospheres with higher-aspect ratios providing more tunable chemical, physical, and biological properties with surface area and energy capable for more interaction with the target analyte or reactive molecules. Various types of nanostructures such as stars, prisms, cubes, circular disks, etc. have been reported by the researchers that are customized as per the required application. If we talk about the effect of nanostructures on LSPR-based sensor sensitivity for different shapes of metal nanostructures, then the nanoshape with maximum number of tips/edges shows higher refractive index sensitivity which can also be confirmed from their aspect ratio. The LSPR spectrum of silver nanoparticles with different shapes: spheres, triangles, and cubes shows that triangular nanostructure records the highest sensitivity (350 nm/RIU) as compared to spherical one (160 nm/RIU) since the triangular nanostructure has the highest aspect ratio (Mock et al. 2003). Similarly, gold nanoshells possess higher sensitivity (409 nm/RIU) as compared to spherical nanoparticles (60 nm/RIU) (Sun and Xia 2002b). However, the sensitivity enhancement for nanocubes and nanospheres has been found to be 2.1×10^8 and 9.7×10^6 times, respectively, for the SERS applications (Rycenga et al. 2011). However, metal nanostars are also one of the most suitable candidates for SERS applications. This is due to the localization of electric field over the tips of the nanostars which provides the extraordinary enhancement in the electric field ($\times 10^{10}$) as compared to other nanostructures, which can also be theoretically evaluated by boundary element method. Colloids are the mixed solutions of various nanostructures and their physical and chemical properties depend on the nanostructure with higher concentration.

4.3.1.3 Magnetic Nanoparticles

Magnetic nanomaterials and nanocomposites are also used for sensing applications. The operation of the nanomaterial relies on the material structure and temperature. Soft and hard nano-magnetic materials made from their bulk form exhibit the enhanced properties resulting in superparamagnetic nature. As the nanomaterial/structure size decreases, an increase in the magnetization and coercivity becomes visible in the magnetization curve. For magnetic nanocomposites, the particles should be just in contact (neither isolated nor desegregated) for the coercive force to be maximum (Filipponi and Sutherland 2012). The force required for demagnetizing a magnet is called coercive force, which is needed to be larger for the magnet to maintain its magnetism in the presence of an opposing magnetic field.

Commonly used magnetic nanoparticles in environmental and biological applications are that of iron, since nickel and cobalt are toxic in nature. Usually, nanoparticles of smaller size, <100 nm, are used due to easier injection and circulation in blood capillaries inside human body without any obstruction. Advantages of using nanoparticles having magnetic nature include facile control of their movement in the capillaries by controlling the magnetic field, lower rates of accumulation, easy dispersion, higher probability to localize near the target, and good diffusion rates. The ability to make the particle active and nonactive when required by providing and removing the magnetic field makes magnetic nanoparticles a better choice in biomedical applications such as nuclear magnetic resonance imaging, drug delivery, and therapeutic activities such as hyperthermia to detect pathogens, cancer therapy, etc. The bio-separation of proteins, cells, and nucleotides can be done by using superparamagnetic nanocolloids which has the ability to control their magnetization. In environmental applications, magnetic nanoparticles are used for the identification of organic and inorganic pollutants employing their adsorption properties (Akbarzadeh et al. 2012).

4.3.2 Nanorods

One of the advantages of utilizing the elongate structures of nanoparticles is the enhanced sensitivity to the change in the external environment because of the higher surface-area-to-volume ratio in comparison to the spherical structure. Moreover, multiplexed system prefers lengthened particles for detection. In the case of luminescence characterization, the aspect ratio affects the luminesce quantum efficiency. The quantum luminesce efficiency depends linearly on square of the aspect ratio. Similarly, the fluorescence also depends linearly on the aspect ratio of nanorods (Burda et al. 2005).

4.3.3 Nanowires

Nanowires have shape that is like nanorods but with a smaller thickness. Usually, the circular structure with aspect ratio <10 is considered as nanorods, while >10 is considered as nanowires. The diameter of nanowire is within the range of 10 nm and has unconstrained length. It has been observed that the nanowires possess less conductivity as compared to the bulk due to the scattering from the boundaries since the mean path of the electron along the length of nanowire is higher than that of its width.

4.3.4 Other Nanostructures

Among other types of structures in nanoparticles, nanotubes (e.g., CNTs) are being utilized because of their flexible nature and electrical properties. The electrical properties of nanotubes depend upon the structural properties of nanotubes such as the number of walls, diameter, and twist. Composites, core shells, and hybrids of nanostructures are also attracting attention these days in optical sensing applications. Core shell nanoparticles have exceeded the potential of nanomaterials by choosing unique combinations, materials, compositions, and approaches (Luo et al. 2009). The materials for core and shell are chosen according to the techniques to be implemented per application. The properties of the nanoparticle depend on the core and outer shell. The structure provides easy immobilization of the bioreceptors to interact with the analyte. The optical methods supporting this technique include spectrophotometry/absorption, SPR/LSPR, and fluorescence. The emission spectrum characteristics such as color, narrowness, symmetry, etc. of the nanoparticles depend on the immobilized receptors and their interaction with analyte. The fluorescence and resonance properties also differ. The interaction between target analyte and bioreceptor changes the refractive index, which is detected by LSPR core shell nanoparticle on which the receptor is immobilized. A metal semiconductor nanocomposite of core shell structure has been reported for monitoring the refractive index variation using SPR technique (Sharma and Gupta 2007). The absorption properties of nanoparticles result in surface-enhanced-absorption-based sensors that can study the specific interactions under infrared spectroscopy. Considering the luminescent properties, the shell helps in enhancing the core's luminescence. By choosing a semiconductor material for shell having larger bandgap than that of the core material, luminescence can be enhanced by minimizing the radiationless recombination, for example, CdSe core with ZnS shell (Dabbousi et al. 1997). There are core shell structures with dielectric shell over metal core, which has application in plasmonic sensors where the resonance band is dependent on the ratio of the core to shell as well on the diameter of the core.

Nanohybrids and nanocomposites with different nanoparticles are prepared by choosing the appropriate combinations for enhancing the desired properties dependent on the determined application. Such combinations provide enhanced luminescence, fluorescence, plasmon resonance, and hence sensitivity. Synthetic polymers with highly branched structure of molecules are called dendrimers. Dendrimers are used in sensing where they are the receptors/sensing functionalities or the matrix for growing other nanostructures for sensing. The structural design of the dendritic molecule can be tuned to generate desired properties of the material and surface interactions. Dendrimers are used as immobilizing adhesives for proteins, enzymes, and nanostructures.

4.4 Nanostructures for Sensing

Metallic and metal oxide nanostructures are gaining interest in engineering and sciences over decades due to their vast potential in nanotechnology. The modified methods of synthesis and manufacturing support easy implementation of nanostructures in conjugation with ligands, enzymes, antibodies, imprinted polymers, and drugs. Nanostructures can be mixed with other materials to make nanocomposites to accomplish a variety of promising applications ranging from daily useful stuffs such as sunscreen to remote sensors in satellites and war missiles. An important development made in the medical field with the emergence of nanotechnology is the usage of magnetic nanoparticles for the diagnosis of tumors and other diseases resulting in the evolution of smart sensors for highly sensitive–selective–quick sensing and biomolecule separation. In addition to sensing applications, these nanoparticles also play a vital role in drug delivery and tissue engineering. The magnetic nanoparticles disperse and diffuse widely and actively in the analyte medium to provide fast detection with better sensitivity and detection limit of the sensor. Another advantage of such type of nanostructures is the reusability. Magneto-optic sensors utilizing the technique of SPR/LSPR have been reported which combine the advantages of magnetic and optical methods. The noble metals supporting plasmonics, in general, exhibit the low magneto-optic activity, while ferromagnetic metals hold best magneto-optic properties with broad resonances. Thus, a combination of these provides magneto-plasmonics for sensing giving high sensitivity to sensors. The supporting properties of orthoferrites ($RFeO_3$) such as VIS-NIR transparency, ferrimagnetism, and good domain wall mobility boost their possibilities in magneto-optic sensing (Arregui 2009).

The influence of physical, electrical, optical, mechanical, and magnetic properties of nanostructures for biomedical and environmental monitoring applications has already been discussed in the previous sections. The interfacial reaction of the nanostructures with the analyte or the enhanced interaction by maneuvering the surfaces of nanostructures leads to the development of sensors with organic/naturally occurring nanostructures. This requires the understanding of the possible interactions of the analyte with the nanostructure. The analyte to be sensed decides the modifications required for both bulk-type and surface-type sensors. In bulk-type sensors, the modification for interaction is done on the surface of nanostructures present within the bulk. In surface-type sensors, the affinity area is created externally by modifying the preexisting surface or by adding a selective and sensitive layer (Kalantar-Zadeh and Fry 2007).

Thus, nanotechnology and nanostructures evolution brought extensive advancements in the field of sensing as a bio/chemical receptor with high-surface area and porous structure, matrix supporting bioreceptor, drug carrier, ligand and layer providing surface interaction enhancement to upgrade the performance of the sensor. The approach or idea of adopting the surface interaction and enhancement depends on the nature of the nanomaterial, structure of the nanomaterial oriented for application, interfacial interactions possible between nanoreceptor/receptor supported with nanomaterial, properties of sample including the stability of chemical reactions and their effects, environment in which device or sensor probe needs to be kept, and the transducer type. The interactions on the surface could be physical, chemical, biological, or combination of these. The force of interaction may vary from the weak type such as noncovalent interaction–adsorption including electrostatic forces (hydrogen bonds, ionic interactions), pi interactions, van der Waals effect (dipole–dipole, dipole–induced dipole, fluctuating dipoles), hydrophobic, encapsulation, intermolecular interactions, etc., to strong type such as covalent bonds, intramolecular interactions, which breaks/forms only in chemical reactions. A brief discussion on the surface interactions in the case of nanomaterials is necessary to properly choose the material and interaction per the analyte to be detected.

Adsorption: The noncovalent interactions including van der Waals, electrostatic/ionic, and hydrophobic effect are categorized under adsorption. Organic nanomaterials such as proteins and enzymes show adsorption properties. The electrostatic immobilization of glucose oxidase on zinc oxide nanorods exploiting the difference in isoelectric points of the glucose oxidase and zinc oxide nanorods has been used for insulinoma detection (Usha et al. 2016). The sensing of profenofos and ascorbic acid making use of molecular-imprinting-based sensors by choosing the weak surface interactions of hydrogen bonding and van der Waals force is also discussed in literature (Shrivastav et al. 2016). Molecular imprinting is a highly selective method for sensing applications that creates tailor-made recognition sites for the analyte to be detected. The method, procedure, implementation, and application of this technique will be discussed in detail in Chapter 6.

Covalent interaction: It is possible to attach either functional groups or cross-linkers on the sensor surface directly to react with either the analyte or with other functional groups. Similarly, functional groups can be attached to the targeted analyte as well. The possibilities of these functional groups to covalently couple to generate carboxylates help in sensing purposes. For example, carboxylic acid is a functional group (–C(=O)–OH) that can covalently couple with other functional groups such as alcohol (single OH dipole), phenol (single OH dipole), amine (NH_2), thiol (single SH dipole), aldehyde (single C=O dipole), ketone (single C=O dipole), etc. The stronger and larger the number of dipoles in the carboxylic acid, the stronger and more hydrogen bonds it supports with other functional groups (Kalantar-Zadeh and Fry 2007). It is also possible to fabricate the surface of the sensor using physical/chemical entrapment and assembling techniques. The detection of molecules with huge size is easily performed by physical entrapment. The membranes are made using polymer and hydrogel in which the molecules of smaller size can easily penetrate. In chemical entrapment, a chemical reaction occurs between the trapped analyte molecule to be detected and the membrane. Solgel materials and methods are used for chemical entrapment of analytes in the matrix. Selective and sensitive layers of sensing with precise thickness and nanomaterial growth are possible to fabricate using self-assembly and layer-by-layer (LBL) assembly. Molecules possessing both hydrophilic and hydrophobic parts are used for self-assembling technique, so that they can attach to sensor surface as well as to the analyte using two parts. This makes the sensor well stable. The technique is used for multianalyte sensing also. A thickness of single layer is attached over sensor surface by dipping the surface in the solution of self-assembling molecules prepared for required time, to give self-assembled monolayer (SAM). An SAM layer exhibiting fluorescent property is reported with distinguishing capabilities of nitroaromatic compound. A multilayered material fabrication over the surface of the sensor utilizing different combinations of materials is possible in LBL assembly. The adsorption of various nanomaterials/molecules can be performed using different functional groups or by electrostatically exploiting the difference in electric charges/isoelectric points. Using the electrostatic attraction between the oppositely charged molecules, the technique of LBL is reported (Decher and Schlenoff 2006). The difference in permeability of different molecules allows the selective separation of analytes. This technique holds application in environmental monitoring as well as in biomedical applications for selective removal of pollutants and separation of biomolecules. Various types of surface interactions using nanomaterials in sensors are shown in Figure 4.10.

4.4.1 Single-Nanoparticle Sensors

A brand-new aspect of nanotechnology in sensing prospect is the single-nanoparticle sensor. The capability of single nanoparticle to detect the adsorption of biomolecule and to analyze the biological and chemical interactions at the molecular level helps to uncover the huge possibilities of nanomaterial in sensing field. The benefits of single-nanoparticle-based detection include smaller analyte volume for detection, multiplexing abilities for large-scale detection, and higher sensitivity with lower limit of detection. For detecting and analyzing the single bio/chemical molecule and to study the interaction, different methods such as fluorescence, optical traps, carbon-based nanomaterials, spectrometry, whispering gallery modes, and plasmonic modes have been used in optical transducer system. Of these, plasmonic single-nanoparticle sensor has gained more attraction due to the

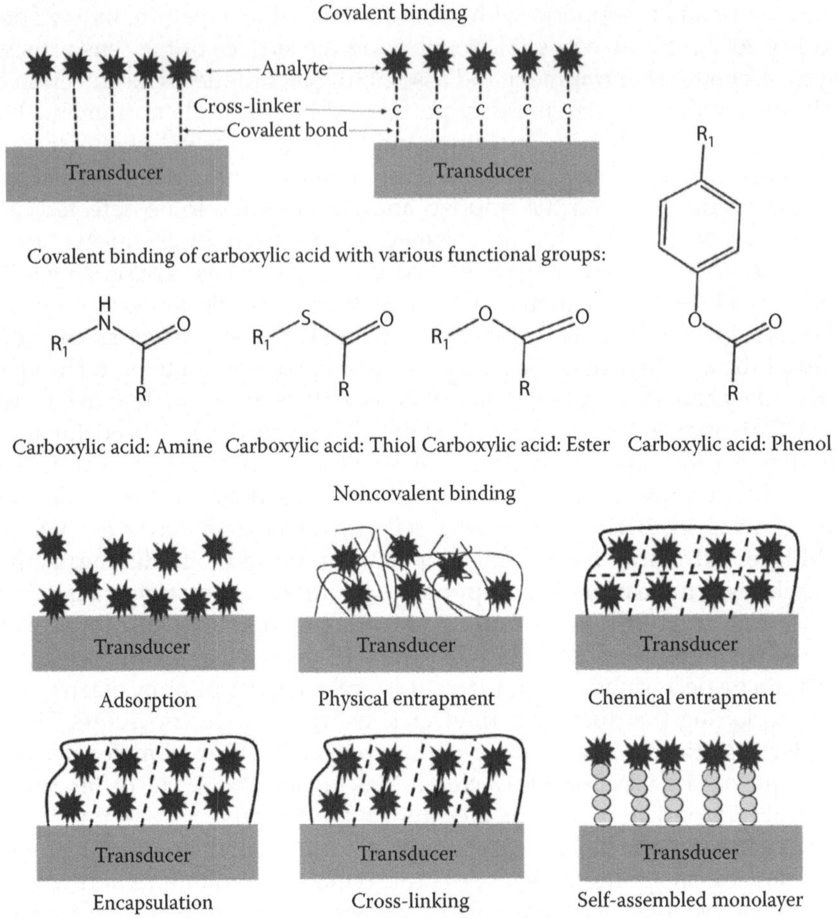

FIGURE 4.10
Possible surface interactions between the transducer surface and the analyte.

ease in localizing the single bioreceptor of extremely smaller size, miniaturized probe realization with point of care application, analyzing the local information on biological cells and tissues for *in vivo* conditions, and implementation of sensor with fast response and reproducibility (Sannomiya and Vörös 2011). Biomarkers, *in vivo* nanosensors, and imaging nanodevices are the possibilities in single nanoparticle system. In plasmonic sensors, the LSPR-based single-nanoparticle sensing system, the sensing at molecular level is performed by means of measuring the shift in resonance/absorbance wavelength. The adsorption of biomolecule on the nanoparticle changes the effective refractive index of the system, which is detected by the shift in resonance/absorbance wavelength. Few examples of single nanoparticle implemented sensor systems are discussed in this section.

An optical sensor for zeptomole designed using a single silver nanoparticle is an example. The sensor based on LSPR technique showed a shift in resonance wavelength of 40.7 nm for a silver nanoparticle that adsorbed around 100 zeptomoles over its surface (McFarland and Van Duyne 2003). Hence, the detection of single biomolecule and analyzing the binding and catalytic reaction rate using single nanoparticle or nanostructure are the considerable developments in the applications of nanotechnology. There are limitations as well, such as the dependence of the sensor response (e.g., in terms of shift in resonance/absorbance wavelength) on the particle size, shape, position, and the distance between particles in multiplexing system. The shortcoming in keeping the size of each nanoparticle identical also affects the sensor system. Regardless of these limitations, the demand for single-nanoparticle-based sensing system in clinical diagnosis and point-of-care applications is increasing even today.

4.4.2 Surface-Enhanced Nanosensors

The development of new nanomaterials, fabrication techniques, instruments for spectroscopic analysis, contemporary detection ideas, and strategy leads to the surface-enhanced nanosensor realization. In the case of plasmonic sensors, the strong signal generation is due to the intense enhancement in electromagnetic fields at the surface of the nanostructures. The surface binding and interaction on real time can be easily studied using SPR/LSPR in which the interaction of the analyte with the sensor surface changes the effective refractive index which shifts the LSPR spectrum with a peak absorbance/resonance wavelength. The main challenge in these plasmonic techniques is the limitation of the sensitivity and limit of detection because of small refractive index variation caused either due to the interaction of analytes with lower molecular weight or due to the lesser quantities of larger-sized analytes. An enhancement in the readable output of such sensors is possible using enzyme-amplified reactor immunoassays, sandwich assays, competitor assays, etc. Immunoassays (antigen–antibody interaction) are used to accurately quantify the concentration of target analyte. The immunoassay that depends on enzymatic reaction to detect/sense analyte is called enzyme immunoassays or enzyme-linked immunosorbent assays. Enzyme amplification immunoassay deals with the usage of an enzyme that generates a catalyst/catalytic promoter for an additional/secondary scheme, which depends on the quantity of enzyme and hence the promoter. The generation of the catalytic promoter and its effect on the secondary scheme results in amplification effect (Self 1985). In sandwich immunoassay, for an antigen, two antibodies are used: a trapping antibody and a measuring/detecting antibody. Among these, trapping antibody is the one specific to antigen interaction (antibody that traps the antigen). In sandwich immunoassay procedure, the trapping antibody is attached to sensor surface followed by the addition of antigen. The detecting antibody is then added which also gets bonded to the antigen to form a sandwich structure. In this method, the quantification is done with the help of indicators such as proteins, enzymes, or fluorophore, which needs to be attached to the detecting antibody or sometimes the detecting antibody itself is the indicator. Thus, as the concentration of antigen/target analyte increases, more detecting antibody gets attached and hence a stronger signal from the indicator gets recorded. The indicator determines the selection of species of trapping and detecting antibodies. If the indicator is not the detecting antibody, then it is possible to select both the antibodies from same species. For the quantitative analysis of small-sized analytes, competitive assay is used. The analyte to be targeted for the detection with competitive assay should be in such a manner that a complemented pair of antibodies is not available for that analyte. Thus, a single antibody is employed for competitive enzyme immunoassay. Enzyme/fluorophore/protein is used as the labeling ligand. The antibody immobilized over the probe surface is incubated with both labeled and unlabeled ligands. As the concentration of target analyte increases, the labeled ligand binding with the analyte increases and hence an enhancement in the sensor response. Figure 4.11 shows the schematic of enzyme amplification, sandwich, and competitive immunoassays for quantification of molecules.

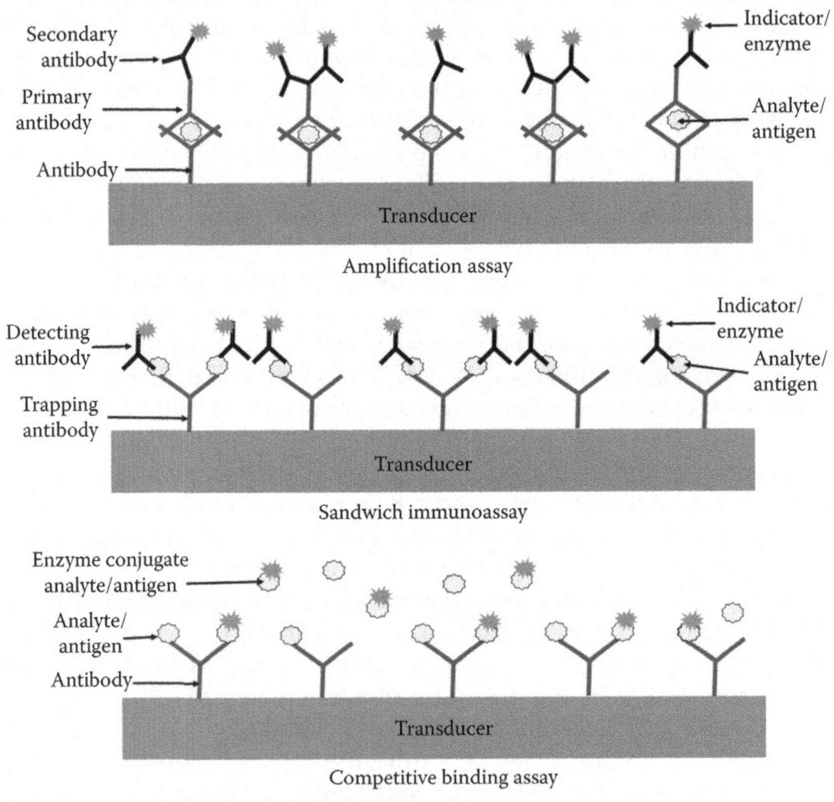

FIGURE 4.11
Three types of immunoassays used for surface-enhanced nanosensing.

The sensitivities of sensors utilizing fluorescence and the Raman scattering can be enhanced by incorporating SPR, which enhances the electromagnetic field (Stewart et al. 2008). Thus, plasmonic nanostructures can enhance the performance of these optical techniques to facilitate many systematic applications. An example of the sensing technique that can be plasmonically enhanced is SERS, where the response signal (and sensitivity of the Raman spectroscopy) is enhanced by 10^6-fold using roughened metallic nanostructures. The high-enhancement factor is because of the electromagnetic and chemical reactive mechanisms since the Raman scattering intensity depends on the induced dipole moment, which is the product of the Raman polarizability and the electromagnetic field intensity of the incident light. Due to the presence of rough metallic nanostructures, localized surface plasmons would be generated on the metallic surface and enhance the electromagnetic field on the surface. The factor of electromagnetic enhancement is 10^4 and that of chemical enhancement is 10^2 (Stiles et al. 2008). SERS sensing holds good sensitivity, selectivity with very negligible interference from water. The usual substrate used for SERS, which is the metal film over nanosphere, is not compatible for stable immobilization because of less affinity of some analytes on metal surfaces. Hence, the substrate needs to be functionalized with groups/layers such as SAM/atomic layer deposition.

In optical sensing, enhancement in the sensor performance is possible using nanoparticles having magnetic properties. Combining magneto-optic sensor nanoparticles with magnetic separators support the detection of the lowest oxygen concentration of 1 mg/L. The magnetic nanoparticles prepared by spray-drying method produced nanoparticles with brightness enhancement of 100 times. The factor needed to be taken care of in this includes the content of magnetite. By applying a magnetic field, the materials can be separated more quickly if the magnetite content in the material is high. At the same time, magnetite possesses good absorption properties which affect the indicator luminescence. Hence, an agreement between the two is recommended (Mistlberger et al. 2009). By regulating the phase of light in magneto-plasmonic nanostructures, an enhancement in sensor's sensitivity and figure of merit can be achieved. Here, LSPR is induced magneto-optically by applying a magnetic field externally. This induces a null point, where the light is linearly polarized at a wavelength, which is capable of accurately determining the change in the light polarization. The phase-dependent LSPR position or peak is identified from the corresponding wavelength. Whenever there is a change in the external refractive index, the evaluation is done by analyzing the shift in the LSPR position/peak.

Thus, in general, if we consider the techniques in optical interrogation strategy for surface enhancement in nanosensing to implement sensor with prominent performance and properties, the premier idea is combining the LSPR with the other methods of optical sensing such as fluorescence, the Raman scattering, magneto-optic, etc. This is because of the great enhancement in the electromagnetic field at the nanostructure surface which can be coupled to the other interfacing method to enhance the performance of the sensor probe. Moreover, the minimal or nanosensing volume requirement, smaller expanse, and higher sensitivity of detection using refractometric setup even for a single molecule identification in a broad variety of liquid medium and systems aid preference to the LSPR-based nanosensors and surface-enhanced sensors in chemical and biomedical sensing applications.

4.5 Applications

The use of nanotechnology in analyzing the physical, chemical, and biological parameters has upgraded the quality and performance of cost-effective, portable sensor devices which play salient roles in improving the standard of the environment in which we live and the healthcare services on which human life is dependent. The compatibility of the nanostructures with the easiest immobilization methods supporting the intricate labeling strategies has made optical sensing of physical, environmental, agricultural, biological, biomedical/clinical variables, and diagnostics facile with precise measurement. In this section, we shall condense the nanostructured optical sensors in four subsections mainly as sensors for physical parameters, environmental and agricultural monitoring, biological applications and biomarkers, and surgical and clinical diagnostics. In addition to the environmental and biomedical nanosensing applications which are the hot topics in the current era of technological development to meet daily needs and industrial practices, the physical parameter analysis also demands close significance. Hence, a brief description of the nanostructured sensors for physical parameter is added in this chapter.

4.5.1 Detection of Physical Parameters

Nanostructured sensors have been reported for the construction sector for monitoring the parameters—pressure, moisture/humidity, temperature, displacement, flow, vibration, corrosion, cracks, shrinkage, viscosity, density of concrete, and so on. The advantages of nanosensors in structural applications involve negligible intrusion with the built-in space and structure, possibility of lodging into the building structure during the construction itself because of the smaller dimensions. A temperature nanosensor using CdTe quantum dots deposited on the tip of the tapered optical fiber head has been reported (de Bastida et al. 2006). The properties of quantum dots to have a narrow emission spectrum and the possibility to change the peak wavelength of the emission spectrum depending on the size of the nanocrystals with the ability to get excited in an extensive range of wavelength make it a fine choice in sensor applications. Quantum dots of sizes 4–5 nm are implemented for temperature monitoring in a range of 30–100°C. The sensor showed a maximum sensitivity of 0.2 nm/°C. A limitation of the design is the photobleaching of the quantum dots which may affect the results. Similar studies have been reported for the humidity detection using optical properties of various nanomaterials which are sensitive toward surrounding atmosphere. Using optical properties of graphene oxide and nano-anatase titanium dioxide, a highly sensitive humidity sensor based on the effect of evanescent field with change in external humidity has been investigated (Ghadiry et al. 2016). Employing the difference in refractive index between graphene oxide and nano-anatase titanium dioxide, relative humidity was evaluated. The part of the EM field propagating through the core escapes from the core as evanescent field and propagates through the cladding. Thus, the field propagating through the core is dependent on the evanescent field as well. In the case of a change in relative humidity, the cladding refractive index changes, which in turn affects the power guided in the core. A general relative humidity sensor works in such a manner that a hike in the relative humidity causes an increase in the water absorption of the sensor coating/sensing layer that increases its refractive index. This changes the propagating field inside the core and results in lesser power drop at the output. However, considering the two materials used for coating: reduced graphene oxide and nano-anatase titanium dioxide, the condition early described is true for titanium dioxide. In the case of reduced graphene oxide, as relative humidity increases, water absorption leads to an increase in the distance between interlayers of graphene oxide. This decreases the refractive index and the power loss at the output and vice versa. Considering the significance of physical parameter measurement in real time on biomedical, environmental, and construction applications, the amount of research carried out (on nanotechnology-implemented physical sensors) so far is markedly less. A manyfold enhancement in the performance can be obtained by modifying the film by nanomaterials with the recent immobilization methods and coupling using suitable receptor for the targeted parameter.

4.5.2 Environmental and Agricultural Monitoring

The steps put forward by the International Environment Union and Committees, for the sake of nature conservation and to minimize the health hazards caused to humans and animals by the waste products and poisonous emissions from industries and the environmental implementations, have grown the public awareness about the human and animal life threat due to the rising environmental issues. Thus, any effectuated environmental application recommends the continual monitoring of the quality of its elemental factors such as water, air, soil, and agricultural products. The efficiency of the designed sensor/tracking system in monitoring the water, atmospheric, and agricultural quality along with regulating the standards of food products plays a major role in ensuring the quality of life. The increasing necessitation of fast and continuous detection of air, water, and soil pollutants/toxic gases in oil, chemical, textile, gas, and agricultural industries shows a great demand for a simple, robust, small-sized, and recoverable sensor, which encourages researchers to design and optimize the novel sensors. Agricultural field demands the evolution of sensors. Sensors are used to ensure the levels of nutrients, pathogens, pesticides, and fertilizer levels in soil. The recent and notable nanostructured sensors reported for the environmental monitoring and agricultural food quality control are discussed in this section.

Environmental monitoring includes atmospheric surveillance and water quality control. Atmospheric parameter monitoring involves detecting the traces of toxic gases and other pollutant exhausts from industries. For instance, methane is an inflammable gas that produces toxic by-products and explosives when it reacts with other pollutants in the atmosphere and is highly harmful to our ecosystem. Although ammonia is used as a component and is produced as a by-product in both chemical and agricultural industries, it is a toxic gas which affects public health. To detect ammonia gas, an SPR-based sensor has been fabricated using a nanocomposite of polymethylmethacrylate and reduced graphene oxide (Mishra et al. 2014). The fabricated sensor has a copper metal layer over unclad core of the optical fiber followed by the nanocomposite layer. The nanocomposite was prepared using *in situ* polymerization technique in which the polymer acts as the matrix for reduced graphene oxide. The doping concentration of reduced graphene oxide in the polymer matrix was 5%. The hydrogen bond formed between ammonia and the oxygen functionalities of the reduced graphene oxide is shown in Figure 4.12.

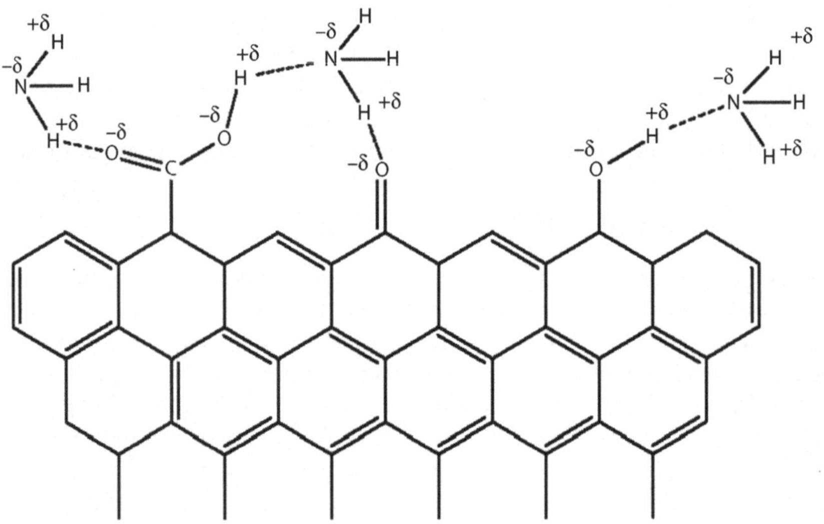

FIGURE 4.12
Hydrogen bond formation between ammonia and oxygen functionalities of reduced graphene oxide. (Reprinted from *Sens. Actuator. B*, 199, Mishra, S. K. et al., SPR based fiber optic ammonia gas sensor utilizing nanocomposite film of PMMA/reduced graphene oxide prepared by in situ polymerization, 190–200. Copyright 2014, with permission from Elsevier.)

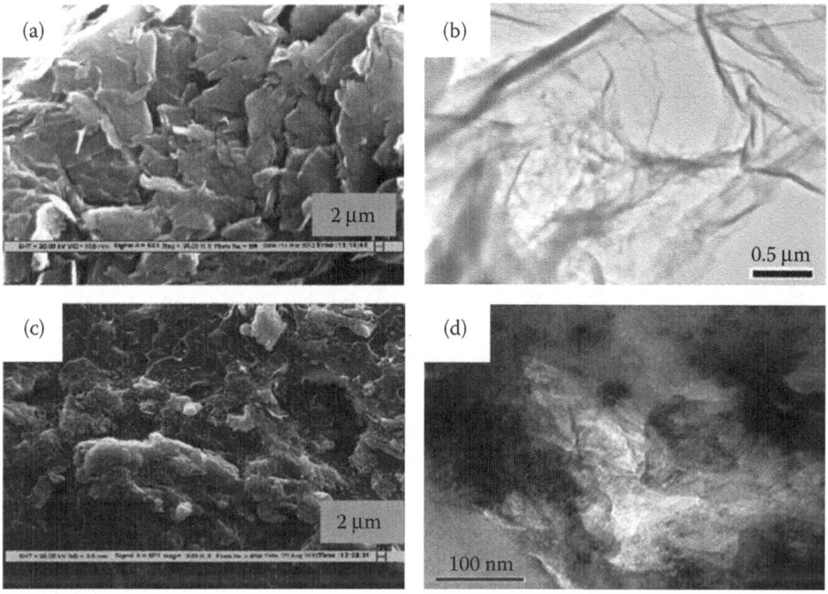

FIGURE 4.13
Morphological SEM and TEM images of (a, b) reduced graphene oxide and (c, d) nanocomposite of polymethylmethacrylate and reduced graphene oxide, respectively. (Reprinted from *Sens. Actuator. B*, 199, Mishra, S. K. et al., SPR based fiber optic ammonia gas sensor utilizing nanocomposite film of PMMA/reduced graphene oxide prepared by in situ polymerization, 190–200. Copyright 2014, with permission from Elsevier.)

The SEM and TEM micrographs of reduced graphene oxide and the nanocomposite with 5% doping concentration are shown in Figure 4.13.

The probe was characterized using the experimental setup shown in Figure 4.14.

The fiber-optic probe was fixed in the gas chamber having the provision for inserting and removing of the gas. The rotary pump and vacuum gauge connected to the chamber ensure the evacuated chamber before each experiment and the existing pressure inside the chamber throughout the experiment, respectively. The light from a tungsten halogen lamp was fed into the fiber and the output light from the fiber was fed into a spectrometer. The reference was recorded when no gas was fed into the chamber. Followed by the reference, ammonia gas was introduced in the gas chamber and the corresponding SPR spectrum was recorded. The reference and the signal were used to determine the SPR spectrum of the gas. The experiments were carried out for 10–100 ppm gas concentration. A 35 nm shift in peak absorbance wavelength was observed for 10–100 ppm gas concentration. The SPR spectra and the calibration curve are shown in Figure 4.15.

A polyhybrid-nanocomposite of reduced graphene oxide, CNTs, and polymethylmethacrylate was used as a sensing layer over silver-coated unclad

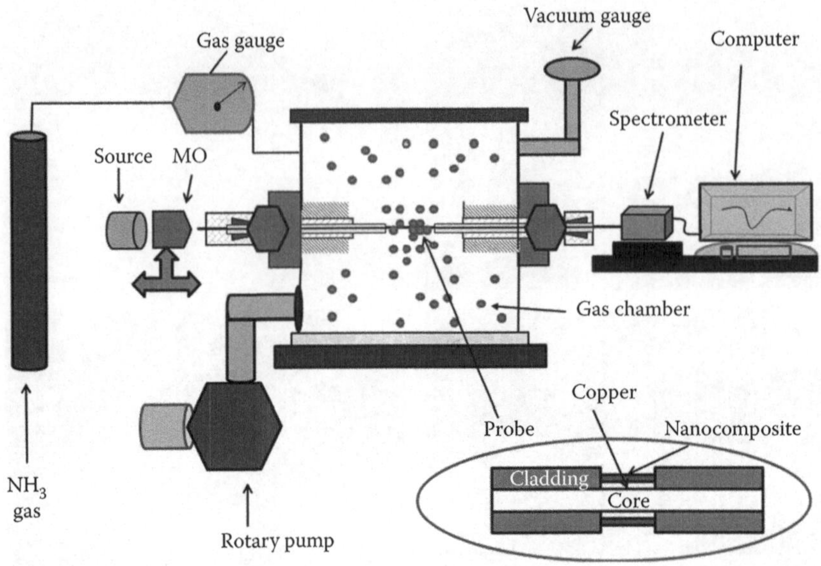

FIGURE 4.14
Optical gas sensing setup used for ammonia sensing. (Reprinted from *Sens. Actuator. B*, 199, Mishra, S. K. et al., SPR based fiber optic ammonia gas sensor utilizing nanocomposite film of PMMA/reduced graphene oxide prepared by in situ polymerization, 190–200. Copyright 2014, with permission from Elsevier.)

core of the optical fiber for methane gas sensing using SPR technique (Mishra et al. 2015). To prepare the nanocomposite, first a hybrid of graphene and CNTs was prepared by *in situ* reduction reaction using hydrazine hydrate. In the second step, nanocomposite of the hybrid and polymethylmethacrylate was prepared employing the sheet casting method and bulk polymerization. The polymer sheets with varying weight percentage of hybrids were prepared and among these the one with a weight percentage of 5 was found to be the optimized doping concentration to give the best performance of the sensor. The structural characterizations of the hybrid and the nanocomposite prepared with 5% weight percentage of hybrid in nanocomposite were carried out using SEM and TEM analysis and are shown in Figure 4.16.

The working of the sensor probe was checked for 10–100 ppm concentration of the methane gas. A total shift of 30 nm in resonance wavelength was obtained for the change in concentration from 10 to 100 ppm. The sensitivity of the probe with other gases (i.e., selectivity) was also checked. Figure 4.17 shows the selectivity test of the probe for various gases.

The nanostructures of semiconducting metal oxides have also been used to design gas sensors. This is because of the distinctive properties of semiconductor metal oxides. The application of semiconductor metal oxides as gas sensors for environmental monitoring will be discussed in detail in Chapter 5.

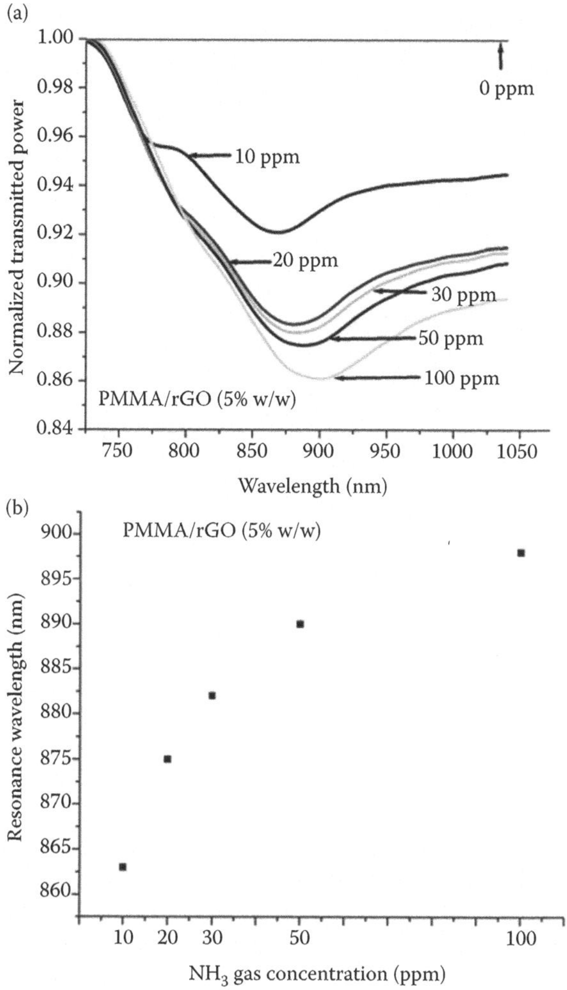

FIGURE 4.15
(a) SPR spectra and (b) the variation of resonance wavelength extracted from SPR spectra with ammonia gas concentration. (Reprinted from *Sens. Actuator. B*, 199, Mishra, S. K. et al., SPR based fiber optic ammonia gas sensor utilizing nanocomposite film of PMMA/reduced graphene oxide prepared by in situ polymerization, 190–200. Copyright 2014, with permission from Elsevier.)

Phenylhydrazine is a pollutant that affects the ecosystem as well as an irritant to human body if exceeds a level. Although hydrazine is an important element in manufacturing drugs, rocket propellants, and pesticides, its residuals found in plants, food items, atmosphere, and even in humans have harmful effects. Recently phenylhydrazine sensor using SPR technique was reported implementing Al@ZnO core shell nanostructured layer over

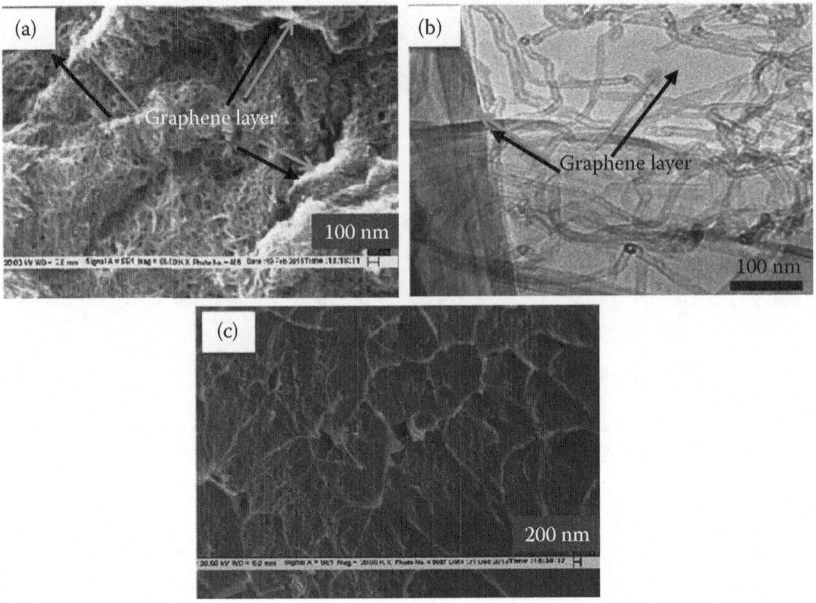

FIGURE 4.16
(a) SEM and (b) TEM micrographs of hybrid having graphene and carbon nanotube, (c) SEM image of nanocomposite formed by hybrid and polymer with 5% hybrid weight percentage. (With kind permission from Springer Science+Business Media: *Plasmonics*, Surface plasmon resonance-based fiber optic methane gas sensor utilizing graphene-carbon nanotubes-poly (methyl methacrylate) hybrid nanocomposite, 10: 2015, 1147–1157, Mishra, S. K. et al.)

silver-film-coated unclad optical fiber core (Tabassum et al. 2016). Laser ablation was used to fabricate the core shell nanostructure. The experimental setup used for the characterization of the biosensor probe is shown in Figure 4.18. The sensor probe fabricated was mounted in the flow cell as shown in the figure. A tungsten halogen lamp was used to launch the light in the fiber and at the other end of the fiber spectrometer was interfaced to a laptop installed with necessary software to analyze the results. The flow cell was having facility of inlet/outlet for easy insertion and removal of analyte without disturbing the setup.

The mechanism of sensing is based on the oxidation of phenylhydrazine due to chemisorbed oxygen in Al@ZnO which results in the formation of diazenyl benzene as intermediate product. The intermediate product again reacts with the chemisorbed oxygen species to form benzenediazonium as the by-product. The formation of intermediate and by-product of the redox reaction change the refractive index of the sensing layer which is recognized by the shift in resonance wavelength. The core shell structure enhances the surface-to-volume ratio, thereby maximizing the sensitivity. The TEM image of core shell nanostructure and the schematic of mechanism of phenylhydrazine are shown in Figure 4.19.

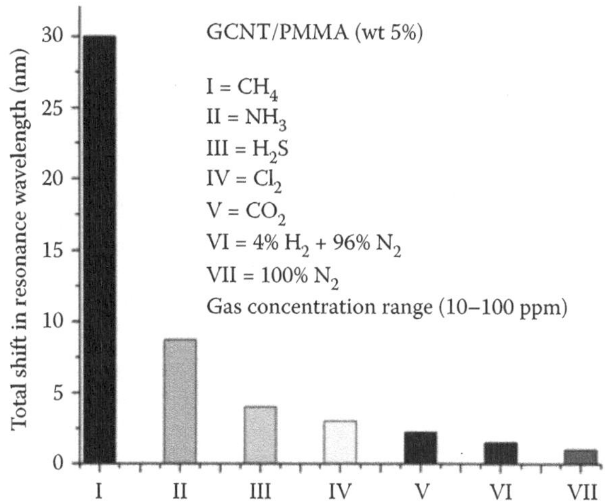

FIGURE 4.17
Response of the probe fabricated using nanocomposite of (graphene–carbon nanotube) hybrid and polymethylmethacrylate for various gases. (With kind permission from Springer Science+Business Media: *Plasmonics*, Surface plasmon resonance-based fiber optic methane gas sensor utilizing graphene-carbon nanotubes-poly (methyl methacrylate) hybrid nanocomposite, 10: 2015, 1147–1157, Mishra, S. K. et al.)

Water quality control involves monitoring of toxins, ions, pathogens, pesticides, persistent organic pollutants (POPs), herbicides, endocrine-disrupting chemicals, and so on. Below, we discuss few nanostructured sensors for the detection of water pollutants.

Persistent organic pollutants: POPs are dangerous category among pollutants reported by Stockholm Convention, which persist for long time

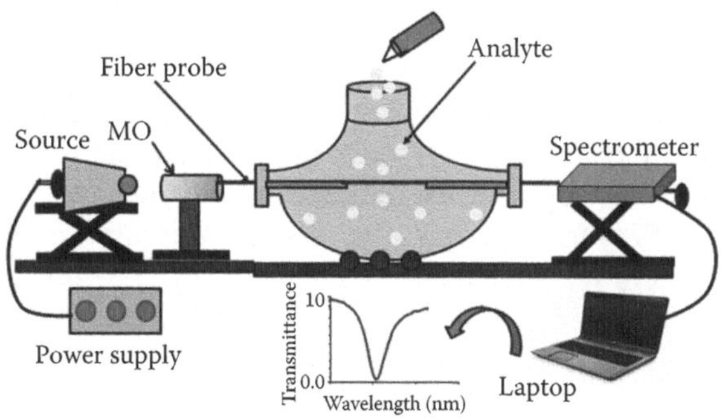

FIGURE 4.18
Schematic of biosensing setup for the detection of phenylhydrazine. (Reprinted from Tabassum, R. et al. 2016. Tuning the field distribution and fabrication of Al@ZnO core shell nanostructure for SPR based fiber optic phenylhydrazine sensor. *Nanotechnology* 27: 215501. With permission from IOP Science.)

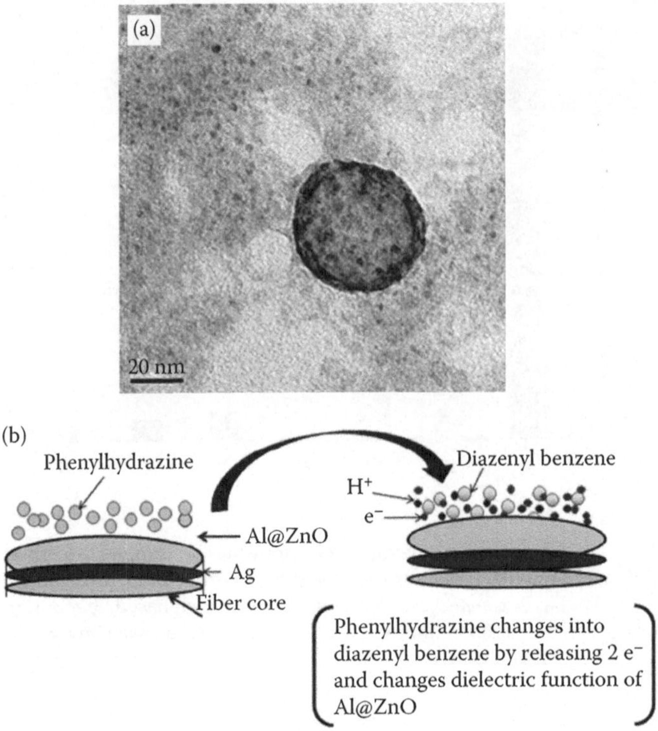

FIGURE 4.19
(a) TEM image of a core shell nanostructure at a magnification of 200 KX and (b) the schematic of sensing principle. (Reprinted from Tabassum, R. et al. 2016. Tuning the field distribution and fabrication of Al@ZnO core shell nanostructure for SPR based fiber optic phenylhydrazine sensor. *Nanotechnology* 27: 215501. With permission from IOP Science.)

and reach animals, aquatic bodies, and plants through soil affecting the ecosystem and human health. Hexachlorobenzene (HCB) is a POP listed under 12 threatening POPs reported by Stockholm Convention. It has been used as a pesticide in agriculture and is ejected as a by-product from several chemical industries. A recent work on industrial application-oriented HCB nanosensor was reported in literature (Sharma et al. 2017). Utilizing the technique of SPR for analysis, the sensor fabricated over unclad core of optical fiber includes silver layer (for realizing SPR) and a nanohybrid of SnO_2 nanoparticles in graphene nanoplatelets (sensing layer). The synergistic behavior of the nanohybrid owed a high sensitivity and selectivity to the sensor. *Ex situ* binding was used to produce the nanohybrid. The pictorial representation of the nanohybrid possibilities—the graphene nanoplatelet surface by SnO_2 nanoparticles/graphene nanoplatelet-encapsulated SnO_2 nanoparticles by the *ex situ* binding between SnO_2 nanoparticles—is shown in Figure 4.20. HCB undergoes dichlorination in the presence of sensing layer which changes the refractive index of the

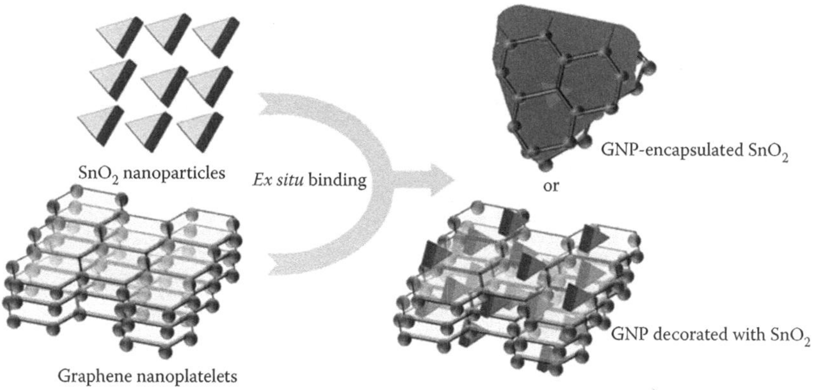

FIGURE 4.20
Ex situ binding possibilities between graphene nanoplatelets and SnO_2 nanoparticles for nano-hybrid formation. (Reprinted from *Sens. Actuator. B*, 246, Sharma, S. et al. novel method of SPR based SnO_2: GNP nano-hybrid decorated optical fiber platform for hexachlorobenzene sensing, 927–936. Copyright 2017, with permission from Elsevier.)

sensing layer. The sensor probe was characterized for HCB concentration range from 0 to 10^{-2} g/L.

Manganese ions: The concentration of any metal ions in its exceeded range than the allowed value would be toxic to ecosystem as well as to the human beings and animals. The Mn^{2+} sensor based on SPR technique utilizing a layer of nanocomposite of ZnO and polypyrrole over silver-coated unclad core of the fiber was reported (Tabassum and Gupta 2015b). The sensor works on the change in the refractive index of the sensing layer due to the interaction of polypyrrole with the manganese ions. The schematic of the probe and the sensing mechanism for manganese ion detection using ZnO:Polypyrrole are shown in Figure 4.21.

The ratio of ZnO and polypyrrole in nanocomposite was optimized and the optimum concentration of polypyrrole in the composite, $ZnO_{(1-x)}$:PPY_x, was obtained as 0.55. The selectivity of the fabricated sensor was checked by experimenting the probe with various metal ions and the maximum sensor response was found for manganese ion, which upholds the practical application of the sensor. The sensor possesses a detection limit of 3.7×10^{-2} ppm, which has been found to be better than the nonnanostructured cross-linked gold–chitosan film-based SPR sensor for the detection of heavy metal ions. However, the nanomaterials are also accompanied with some health-related issues and are phytotoxic depending on the choice of nanomaterial since different nanomaterials exhibit different effects on the plants and animals. The current techniques and application criteria should be developed to enhance the efficiency, biodegradability–biocompatibility of nanomaterials and structures on the analyte required to be monitored, and more over to make it environment-friendly.

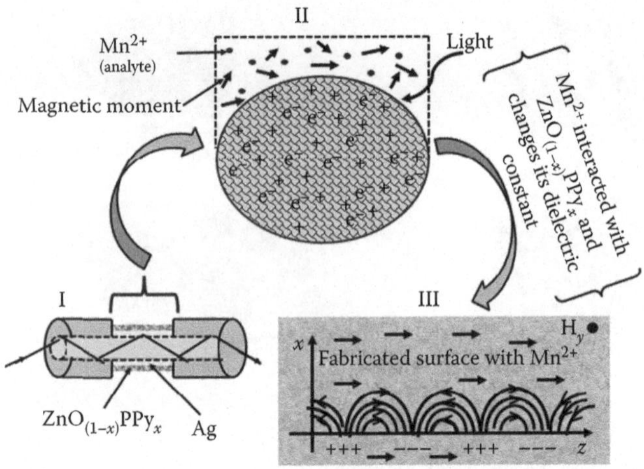

FIGURE 4.21
Mechanism of sensing of manganese ion using ZnO:Polypyrrole nanocomposite in SPR-based sensing system. (Reprinted from *Sens. Actuator. B*, 220, Tabassum, R. and B. D. Gupta, Fiber optic manganese ions sensor using SPR and nanocomposite of ZnO–Polypyrrole, 903–909. Copyright 2015, with permission from Elsevier.)

4.5.3 Biological Applications and Biomarkers

A reliable method for examining various chemicals and biomolecules in nanoscopic and microscopic environment has been developed using nanotechnology and nanosensors, especially implemented using optical methods. The advantages of optical sensing method of nanotechnological realization include visualizing sensing effect using fluorescence or calorimetric detection, instantaneous response with highly sensitive and selective probe designed with SPR/LSPR/LMR technique, enhanced performance using SERS, point-of-care sensing with online monitoring, real-time applications, and so on.

Optical nanosensors also play an important role in intracellular measurements. An optical fiber nanosensor for the ultrasensitive detection of glutamate by immobilizing the glutamate dehydrogenase enzyme is an example (Cordek et al. 1999). For probe fabrication, an immobilizing method of covalent binding was used. The optical fiber surface was initially provided with an aminosilane coating. Glutaraldehyde was used as a linker that binds covalently the aminosilane layer and the glutamate dehydrogenase. It also acts as a spacer layer that ensures the better accessibility and arrangement of enzyme. The enzyme gets attached on the CHO group of glutaraldehyde. The glutamate detection is carried out by analyzing the fluorescence of NADH, formed by the reaction between NAD^+ and glutamate. Combining design modification of the optical substrate with the technique of LSPR brought up a novel sensor probe for glucose detection (Srivastava et al. 2012). The sensor has been outlined as a U-shaped probe with high sensitivity and

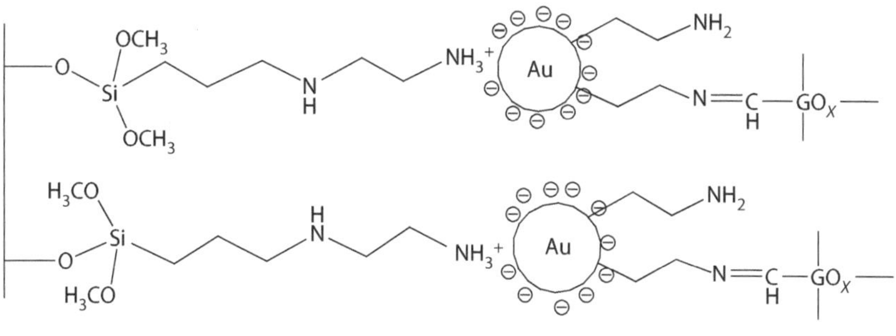

FIGURE 4.22
Self-assembled monolayer used for attaching gold nanoparticle over fiber core and glucose oxidase immobilized over the nanoparticles using cystamine. (With kind permission from Springer Science+Business Media: *Plasmonics*, Localized surface plasmon resonance based fiber optic U-shaped biosensor for the detection of blood glucose. 7: 2012, 261–268, Srivastava, S. K. et al.)

point-of-care application. Over the U-shaped probe made on unclad portion of optical fiber, gold nanoparticles were immobilized using SAM layer of aminosilane. Over the gold nanoparticles, $-NH_2$ group was created by dipping gold nanoparticle-coated fiber in cystamine dihydrochloride. Similarly, the glucose oxidase enzyme, the receptor, was also modified to convert the carbohydrate group to aldehydes. After separate treatment of probe with gold nanoparticles and glucose oxidase, the enzyme was immobilized on the probe surface. The self-assembled monolayer over the unclad fiber core probe is shown in Figure 4.22.

The sensor was characterized by recording the absorbance spectra for glucose concentration range of 0–250 mg/dL and the absorbance spectra obtained for the probe with a bending radius of 0.711 mm of the U-region is shown in Figure 4.23. A decrement in the absorbance value was observed with an increment in the glucose concentration which upholds the metal nanoparticle's polarization variation which in turn changes the scattering and absorption properties.

A very recent study performed using LMR exhibited by zinc oxide showed a novel hydrogen peroxide biosensor with a twin regime of operation (Usha et al. 2017). The operation of sensor supports both *in vivo* applications and hydrogen peroxide liberation range from polyphenolic beverages (such as tea and coffee) under pseudophysiological conditions. The *in vivo* range covers the needed, semitoxic and toxic concentration levels of hydrogen peroxide in human body. The schematic of the sensor probe used for the sensing of hydrogen peroxide is shown in Figure 4.24.

Over the unclad core of optical fiber, a zinc oxide thin film was coated to realize LMR. To grow nanorods over the film, nanoparticles of zinc oxide were coated to act as seed layer to grow nanorods of zinc oxide. The growth period of zinc oxide nanorods was optimized to get the best performance of the sensor. Silver nanoparticles were grown photochemically from silver

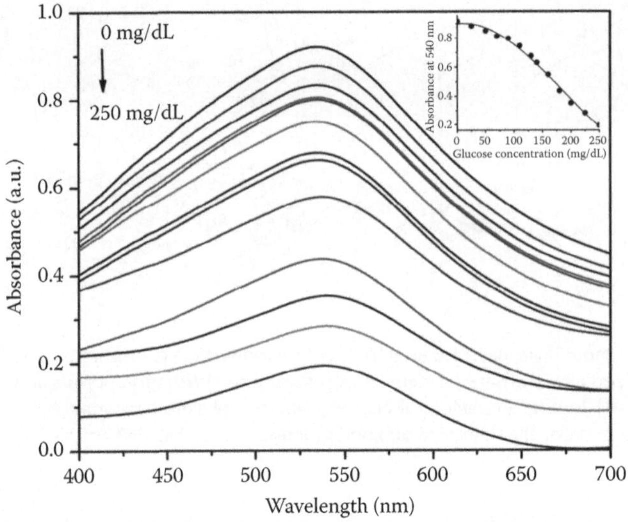

FIGURE 4.23
LSPR spectra of U-shaped glucose sensor for different glucose concentrations. (With kind permission from Springer Science+Business Media: *Plasmonics*, Localized surface plasmon resonance based fiber optic U-shaped biosensor for the detection of blood glucose. 7: 2012, 261–268, Srivastava, S. K. et al.)

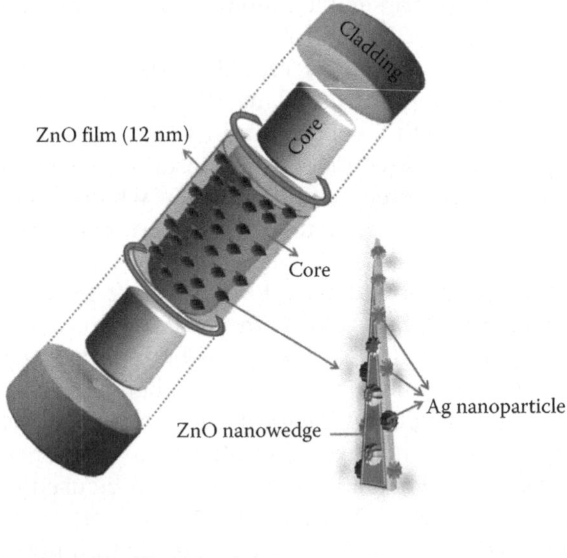

FIGURE 4.24
Pictorial representation of the LMR-based sensor probe designed for hydrogen peroxide sensing. (Reprinted from *Sens. Actuator. B*, 241, Usha, S. P. et al., Silver nanoparticle noduled ZnO nanowedge fetched novel FO-LMR based H_2O_2 biosensor: A twin regime sensor for in-vivo applications and H_2O_2 generation analysis from polyphenolic daily devouring beverages, 129–145. Copyright 2017, with permission from Elsevier.)

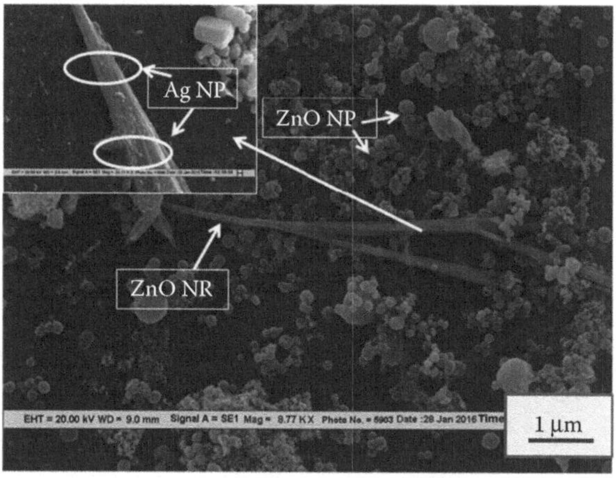

FIGURE 4.25
SEM morphological image of the zinc oxide nanowedges and the nodule formation of silver nanoparticles on zinc oxide nanowedges. (Reprinted from *Sens. Actuator. B*, 241, Usha, S. P. et al., Silver nanoparticle noduled ZnO nanowedge fetched novel FO-LMR based H_2O_2 biosensor: A twin regime sensor for in-vivo applications and H_2O_2 generation analysis from polyphenolic daily devouring beverages, 129–145. Copyright 2017, with permission from Elsevier.)

nitrate solution having probe with zinc oxide nanorods. The probe was irradiated for morphological evolution of nanorods to scaled structure of nanowedges along with the growth of the silver nanoparticles on its surface. The SEM micrograph of the zinc oxide nanowedges with silver nanoparticles is shown in Figure 4.25.

Zinc oxide helps in the generation of silver nanoparticles on its surface by its photocatalytic property. The LMR response curves obtained for hydrogen peroxide concentration of 0–1500 μM are shown in Figure 4.26a along with the calibration curve where the concentration up to 100 μM shows the *in vivo* application range and concentration >100 μM shows the polyphenolic beverage induced hydrogen peroxide generation (Figure 4.26b).

The sensor works on the change in the refractive index due to the series of reactions involved between hydrogen peroxide and sensing layer which causes the change in the peak absorbance wavelength of the LMR spectrum. Depending on the nature of the LMR spectrum obtained for each concentration in the total range considered, the analysis span was divided from 0 to 300 μM as regime 1 and 300 to 1500 μM as regime 2. The method showed its application for the detection of hydrogen peroxide from polyphenolic beverages such as green tea, coffee, and black tea as well.

C-reactive protein (CRP) is a well-known biomarker for infections and inflammatory diseases in human blood serum. A CRP detection method based on SPR phenomenon was reported (Meyer et al. 2006). The sensor was fabricated over a glass chip substrate. CRP antibodies were immobilized over the gold surface, coated on the glass chip, using SAM. The interaction of

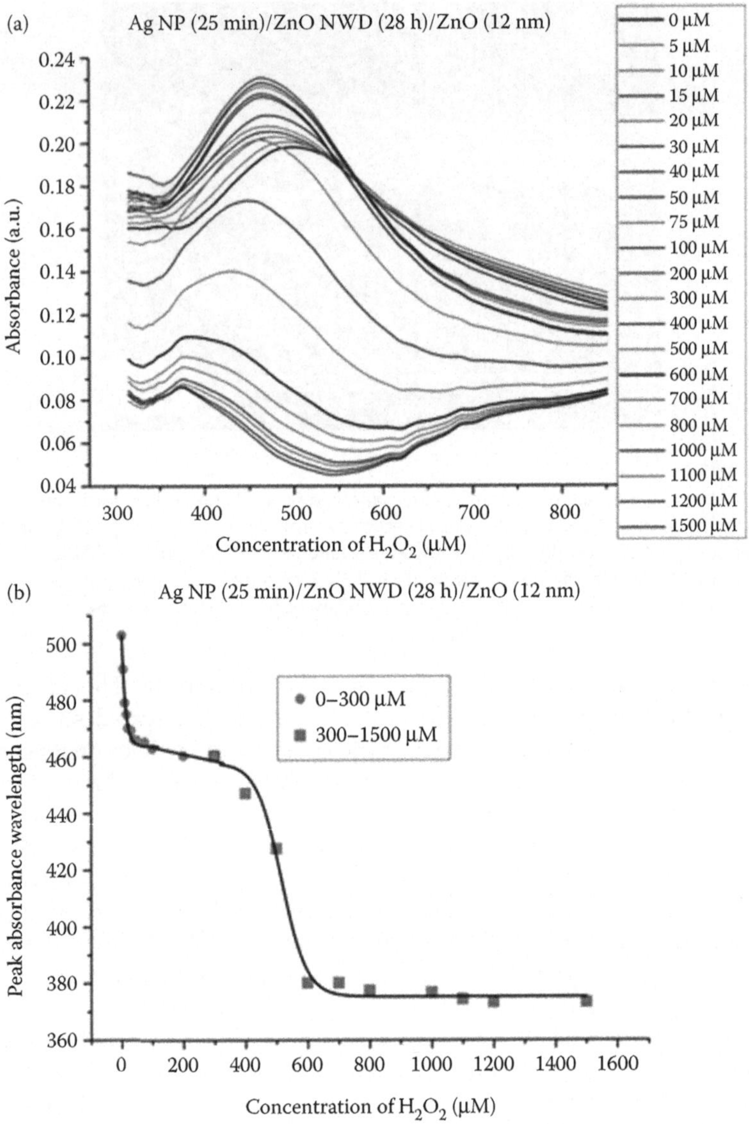

FIGURE 4.26
(a) LMR spectra and (b) calibration curve for the hydrogen peroxide LMR sensor. (Reprinted from *Sens. Actuator. B*, 241, Usha, S. P. et al., Silver nanoparticle noduled ZnO nanowedge fetched novel FO-LMR based H_2O_2 biosensor: A twin regime sensor for in-vivo applications and H_2O_2 generation analysis from polyphenolic daily devouring beverages, 129–145. Copyright 2017, with permission from Elsevier.)

CRP antigens with antibodies results in the change in the effective refractive index of the sensing layer which is deflected in the SPR spectrum as a shift in the resonance angle.

4.5.4 Surgical and Clinical Diagnostics

For highly sensitive and selective detection of diseases, hormones, and surgical conditions, nanosensor-based bio-schemes are currently in the top priority list of tools and devices for surgeons and lab technicians. Few of the advances made by nanotechnology in biomedical applications are discussed in this section.

Subcellular analyte sensing/monitoring is an interesting and sensitive area where nanosystems have already made their mark. A probe encapsulated by biologically localized embedding (PEBBLE) nano-biosensor was reported for subcellular measurements (Clark et al. 1998). The work claimed to bring up the "world's smallest stand-alone devices/sensors," which were composed of fluorescein, matrix, and multiple components—about seven—that include nanospherical structures of diameter 20 nm. The sensor probe, which holds only 1 ppb space in a cell volume of mammals, was reported to be highly selective and reproducible with considerable stability. The designed probe could analyze oxygen, pH, potassium, calcium, and magnesium in cells. The advantages of PEBBLE nanosensor include biocompatibility, no fluorescence quenching, minimal damage to cells, higher signal-to-noise ratio, etc.

The detection of virus, bacteria, influenza, and infectious diseases is essential as they are contagious and it is difficult to control their transfer in a short period. Utilizing SPR technique, a sensor probe with dual channel for simultaneous detection of heparin and vitamin K_1 was designed (Tabassum and Gupta 2016). The schematic of the sensor probe is shown in Figure 4.27, where channel 1 has a coating of silver layer followed by a nanohybrid of CNT in chitosan for vitamin K_1 detection, while channel 2 has a coating of copper over which a nanofilm of polybrene@ZnO core shell structure was given for heparin detection.

Estimation of vitamin K_1 is necessary for monitoring the coagulation of blood while heparin is an anticoagulant. The deficiency and excess of vitamin K_1 are not recommended in human body. Coagulants such as heparin are used to avoid blood coagulation by thrombocytopenia, which is used mostly in surgery. The interaction of the vitamin K_1 with the nanohybrid results in a redox reaction, which changes the refractive index of the sensing layer around channel 1. At the same time, the neutralization of heparin occurs in the presence of polybrene, which is the core enveloped by ZnO shell structure, that changes the refractive index of the sensing layer around channel 2. The SPR spectra recorded for the mixture of vitamin K_1 and heparin with equal concentrations in the range from 0 to 10^{-3} g/L in each sample are shown in Figure 4.28. The well-separated SPR dips in the response curve show the sustainability of the chosen materials and their combinations. The

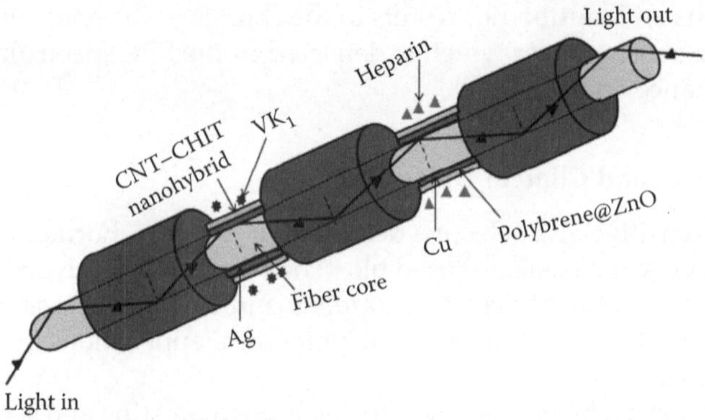

FIGURE 4.27
Schematic of dual-channel SPR-based probe for vitamin K_1 and heparin detection. (Reprinted from *Biosens. Bioelectron.*, 86, Tabassum, R. and B. D. Gupta, Simultaneous estimation of vitamin K_1 and heparin with low limit of detection using cascaded channels fiber optic surface plasmon resonance, 48–55. Copyright 2016, with permission from Elsevier.)

work revealed a cascaded approach useful in medical diagnostics, where both the deficiency and excess of two separate targets can be measured.

Cytokines are the bioactive proteins responsible for the regulation of the immune cells in human body. The protein monitors the responsiveness, maturation, and the growth of immune cells. The detection of cytokines in human serum provides highly valuable clinical information for monitoring

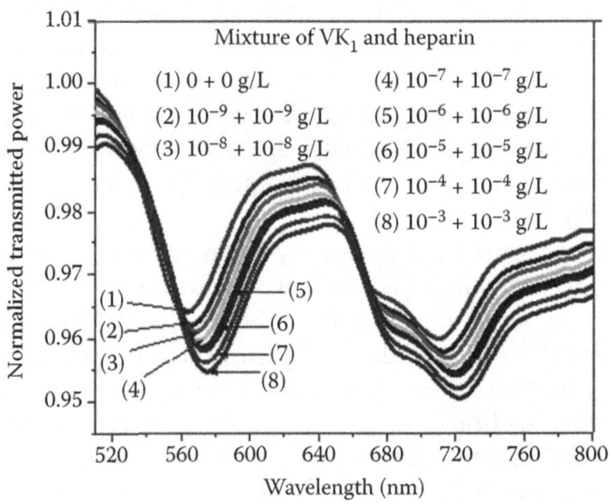

FIGURE 4.28
SPR spectra showing the simultaneous detection of vitamin K_1 and heparin using dual-channel probe. (Reprinted from *Biosens. Bioelectron.*, 86, Tabassum, R. and B. D. Gupta, Simultaneous estimation of vitamin K_1 and heparin with low limit of detection using cascaded channels fiber optic surface plasmon resonance, 48–55. Copyright 2016, with permission from Elsevier.)

the immune status of a person. Hence, the protein is, generally, monitored before and after the surgery in patients suffering from different inflammatory diseases such as cancer, sepsis, and lupus. An LSPR-based biosensor for the detection of cytokines concentrations in human blood samples was reported (Chen et al. 2015). Microarrays of cytokines antibody-functionalized gold nanorods were integrated on a glass chip. The binding of cytokine, in blood samples, with the immobilized antibodies causes the change in refractive index of the medium surrounding the gold nanorods. A shift in the absorbance wavelength observed from LSPR spectrum before and after binding of cytokines confirms the change in concentration of cytokine present. Cardiopulmonary bypass (CPB) is a machine that mimics the function of heart and lungs during an open-heart surgery. The designed sensor has been used along with CPB to monitor the cytokine levels in blood samples.

The applications of nanotechnology and nanomaterials in biological, medical, surgical, and clinical ways are emerging gradually. The use of nanomaterials to design biosensor probes provides good response, high sensitivity, selectivity, shorter response time, better reproducibility, and recovery. The advantages of nanomaterials such as direct adsorption of biomolecules over bioreceptors and the ability to tune their properties and response for an entailed application make them a relevant choice for device implementation.

4.6 Summary

In this chapter, we have discussed sensors based on various kinds of nanostructures. The chapter begins with nanotechnology as a sensing platform and discusses physical, chemical, and biological methods of the synthesis of the nanostructures of different shapes. Emphasis has been given on gold, silver, platinum, and palladium metals. Single and surface-enhanced nanosensors for the detection of different kinds of analytes have been presented.

References

Akbarzadeh, A., M. Samiei, and S. Davaran. 2012. Magnetic nanoparticles: Preparation, physical properties, and applications in biomedicine. *Nanoscale Res. Lett.* 7: 144.

Arregui, F. J. 2009. *Sensors Based on Nanostructured Materials*. New York: Springer.

Aslan, K., Z. Leonenko, J. R. Lakowicz, and C. D. Geddes. 2005. Fast and slow deposition of silver nanorods on planar surfaces: Application to metal-enhanced fluorescence. *J. Phys. Chem. B* 109: 3157–3162.

Bera, D., S. C. Kuiry, M. McCutchen et al. 2004. *In-situ* synthesis of palladium nanoparticles-filled carbon nanotubes using arc-discharge in solution. *Chem. Phys. Lett.* 386: 364–368.

Boutinguiza, M., M. Meixus, J. del Val et al. 2016. Synthesis and characterization of Pd nanoparticles by laser ablation in water using nanosecond laser. *Phys. Procedia.* 83: 36–45.

Brust, M., M. Walker, D. Bethell, D. J. Schiffrin, and R. Whyman. 1994. Synthesis of thiol-derivatised gold nanoparticles in a two-phase liquid-liquid system. *Chem. Commun.* 7: 801–802.

Burda, C., X. Chen, R. Narayanan, and M. A. El-Sayed. 2005. Chemistry and properties of nanocrystals of different shapes. *Chem. Rev.* 105: 1025–1102.

Cai, W., T. Gao, H. Hong, and J. Sun. 2008. Application of Au nanoparticles in cancer nanotechnology. *Nanotech. Sci. Appl.* 1: 17–32.

Campbell, S. A. 2001. *The Science and Engineering of Microelectronic Fabrication.* New York: Oxford University Press.

Castro-Longoria, E., S. D. Moreno-Velázquez, A. R. Vilchis-Nestor, E. A. Berumen, and M. A. Borja. 2012. Production of platinum nanoparticles and nanoaggregates using *Neurospora crassa. J. Microbiol. Biotechnol.* 22: 1000–1004.

Chen, P., M. T. Chung, W. McHugh et al. 2015. Multiplex serum cytokine immunoassay using nanoplasmonic biosensor microarrays. *ACS Nano* 9: 4173–4181.

Chimene, D., D. L. Alge, and A. K. Gahawar. 2015. Two-dimensional nanomaterials for biomedical applications: Emerging trends and future prospects. *Adv. Mater.* 27: 7261–7284.

Clark, H. A., S. L. Barker, M. Brasuel et al. 1998. Subcellular optochemical nanobiosensors: Probes encapsulated by biologically localised embedding (PEBBLEs). *Sens. Actuators B* 51: 12–16.

Cliffel, D. E., F. P. Zamborini, S. M. Gross, and R. W. Murray. 2000. Mercaptoammonium-monolayer-protected, water-soluble gold, silver, and palladium clusters. *Langmuir* 16: 9699–9702.

Cordek, J., X. Wang, and W. Tan. 1999. Direct immobilization of glutamate dehydrogenase on optical fiber probes for ultrasensitive glutamate detection. *Anal. Chem.* 71: 1529–1533.

Dabbousi, B. O., J. Rodriguez-Viejo, F. V. Mikulec et al. 1997. (CdSe) ZnS core–shell quantum dots: Synthesis and characterization of a size series of highly luminescent nanocrystallites. *J. Phys. Chem. B* 101: 9463–9475.

de Bastida, G., F. Arregui, J. Goicoechea, and I. R. Matias. 2006. Quantum dots-based optical fiber temperature sensors fabricated by layer-by-layer. *IEEE Sens. J.* 6: 1378–1379.

Decher, G. and J. B. Schlenoff. 2006. *Multilayer Thin Films: Sequential Assembly of Nanocomposite Materials.* United States: John Wiley.

Fayaz, A. M., K. Balaji, M. Girilal, R. Yadav, P. T. Kalaichelvan, and R. Venketesan. 2010. Biogenic synthesis of silver nanoparticles and their synergistic effect with antibiotics: A study against gram-positive and gram-negative bacteria. *Nanomed. Nanotechnol.* 6: 103–109.

Filipponi, L. and D. Sutherland. 2012. *Nanotechnologies: Principles, Applications, Implications and Hands-On Activities.* Luxembourg: European Commission.

Gaspar, D. P., A. C. Pimentel, T. Mateus et al. 2013. Influence of the layer thickness in plasmonic gold nanoparticles produced by thermal evaporation. *Sci. Rep.* 3: 1469.

Ghadiry, M., M. Gholami, C. K. Lai, H. Ahmad, and W. Y. Chong. 2016. Ultrasensitive humidity sensor based on optical properties of graphene oxide and nano-anatase TiO_2. *PloS One* 11: 0153949.

Govindaraju, K., S. K. Basha, V. G. Kumar, and G. Singaravelu. 2008. Silver, gold, and bimetallic nanoparticles production using single cell protein (*Spirulina platensis*) geitler. *J. Mater. Sci.* 43: 5115–5122.

Gu, X., C. Nie, Y. Lai, and C. Lin. 2006. Synthesis of silver nanorods and nanowires by tartrate-reduced route in aqueous solutions. *Mater. Chem. Phys.* 96: 217–222.

Haes, A. J. and R. P. Van Duyne. 2003. Nanosensors enable portable detectors for environmental and medical applications. *Laser Focus World* 39: 153–156.

Hu, J. Q., Q. Chen, Z. X. Xie et al. 2004. A simple and effective route for the synthesis of crystalline silver nanorods and nanowires. *Adv. Funct. Mater.* 14: 183–189.

Hulla, J. E., S. C. Sahu, and A. W. Hayes. 2015. Nanotechnology: History and future. *Hum. Exp. Toxicol.* 34: 1318–1321.

Humphrey, S. M., M. E. Grass, S. E. Habas, K. Niesz, G. A. Somorjai, and T. D. Tilley. 2007. Rhodium nanoparticles from cluster seeds: Control of size and shape by precursor addition rate. *Nano Lett.* 7: 785–790.

Iravani, S. 2011. Green synthesis of metal nanoparticles using plants. *Green Chem.* 13: 2638–2650.

Islam, M. A., M. A. K. Bhuiya, and M. S. Islam. 2014. A review on chemical synthesis process of platinum nanoparticles. *Asia Pac. J. Energy Environ.* 1: 107–120.

Jan, T., J. Iqbal, M. Ismail et al. 2014. Synthesis, physical properties and antibacterial activity of metal oxides nanostructures. *Mater. Sci. Semicond. Process* 21: 154–160.

Jana, N. R., L. Gearheart, and C. J. Murphy. 2001. Seed-mediated growth approach for shape-controlled synthesis of spheroidal and rod-like gold nanoparticles using a surfactant template. *Adv. Mater.* 13: 1389–1393.

Jeff, R. C., M. Yun, B. Ramalingam et al. 2011. Charge storage characteristics of ultra-small Pt nanoparticle embedded gaAs based non-volatile memory. *Appl. Phys. Lett.* 99: 072104.

Kalantar-Zadeh, K. and B. Fry. 2007. *Nanotechnology-Enabled Sensors*. Germany: Springer.

Kelly, K. L., E. Coronado, L. L. Zhao, and G. C. Schatz. 2003. The optical properties of metal nanoparticles: The influence of size, shape, and dielectric environment. *J. Phys. Chem. B* 107: 668–677.

Khaydarov, R. A., R. R. Khaydarov, O. Gapurova, Y. Estrin, and T. Scheper. 2009. Electrochemical method for the synthesis of silver nanoparticles. *J. Nanopart. Res.* 11: 1193–1200.

Kimling, J., M. Maier, B. Okenve, V. Kotaidis, H. Ballot, and A. Plech. 2006. Turkevich method for gold nanoparticle synthesis revisited. *J. Phys. Chem. B* 110: 15700–15707.

Kruis, F., H. Fissan, and B. Rellinghaus. 2000. Sintering and evaporation characteristics of gas phase synthesis of size selected PbS nanoparticles. *Mater. Sci. Eng. B* 69: 329–334.

Krutyakov, Y. A., A. Y. Olenin, A. A. Kudrinskii, P. S. Dzhurik, and G. V. Lisichkin. 2008. Aggregative stability and polydispersity of silver nanoparticles prepared using two-phase aqueous organic systems. *Nanotechnol. Russ.* 3: 303–310.

Kumar, P. S., I. P. Santos, B. R. González, F. J. G. Abajo, and L. M. Liz-Marzán. 2007. High-yield synthesis and optical response of gold nanostars. *Nanotechnology* 19: 015606.

Kumar, S. A., M. K. Abyaneh, S. W. Gosavi et al. 2007. Nitrate reductase-mediated synthesis of silver nanoparticles from $AgNO_3$. *Biotechnol. Lett.* 29: 439–445.

LaMer, V. K. 1952. Nucleation in phase transitions. *Ind. Eng. Chem.* 44: 1270–1277.

Lim, P. Y., R. S. Liu, P. L. She, C. F. Hung, and H. C. Shih. 2006. Synthesis of Ag nanospheres particles in ethylene glycol by electrochemical-assisted polyol process. *Chem. Phys. Lett.* 420: 304–308.

Liu, S., J. Yue, and A. Gedanken. 2001. Synthesis of long silver nanowires from AgBr nanocrystals. *Adv. Mater.* 13: 656–658.

Logeeswaran, V. J., N. P. Kobayashi, M. S. Islam et al. 2008. Ultrasmooth silver thin films deposited with a germanium nucleation layer. *Nano Lett.* 9: 178–182.

Lung, J. K., J. C. Huang, D. C. Tien et al. 2007. Preparation of gold nanoparticles by arc discharge in water. *J. Alloy Compd.* 434: 655–658.

Luo, J., G. Zeng, L. Tang, J. Yin, and Y. Li. 2009. Study progress on biosensing core/shell nanoparticles. *Chinese J. Anal. Chem.* 37: 1847–1852.

Mafuné, F., J. Kohno, Y. Takeda, and T. Kondow. 2002. Full physical preparation of size-selected gold nanoparticles in solution: Laser ablation and laser-induced size control. *J. Phys. Chem. B* 106: 7575–7577.

Mafuné, F., J. Kohno, Y. Takeda, T. Kondow, and H. Sawabe. 2000a. Structure and stability of silver nanoparticles in aqueous solution produced by laser ablation. *J. Phys. Chem. B* 104: 8333–8337.

Mafuné, F., J. Kohno, Y. Takeda, T. Kondow, and H. Sawabe. 2000b. Formation and size control of silver nanoparticles by laser ablation in aqueous solution. *J. Phys. Chem. B* 104: 9111–9117.

Magnusson, M. H., K. Deppert, J. O. Malm, J. O. Bovin, and L. Samuelson. 1999. Gold nanoparticles: Production, reshaping, and thermal charging. *J. Nanopart. Res.* 1: 243–251.

Malsch, I. 2002. Biomedical applications of nanotechnology. *Industrial Physicist* 8: 15–17.

Manna, A., P. L. Chen, H. Akiyama, T. X. Wei, K. Tamada, and W. Knoll. 2003. Optimized photoisomerization on gold nanoparticles capped by unsymmetrical azobenzene disulfides. *Chem. Mater.* 15: 20–28.

McFarland, A. D. and R. P. Van Duyne. 2003. Single silver nanoparticles as real-time optical sensors with zeptomole sensitivity. *Nano Lett.* 3: 1057–1062.

Meyer, M. H. F., M. Hartmann, and M. Keusgen. 2006. SPR-based immunosensor for the CRP detection—A new method to detect a well known protein. *Biosens. Bioelectron.* 21: 1987–1990.

Mishra, S. K., S. N. Tripathi, V. Choudhary, and B. D. Gupta. 2014. SPR based fiber optic ammonia gas sensor utilizing nanocomposite film of PMMA/reduced graphene oxide prepared by in situ polymerization. *Sens. Actuator. B* 199: 190–200.

Mishra, S. K., S. N. Tripathi, V. Choudhary, and B. D. Gupta. 2015. Surface plasmon resonance-based fiber optic methane gas sensor utilizing graphene-carbon nanotubes-poly (methyl methacrylate) hybrid nanocomposite. *Plasmonics* 10: 1147–1157.

Mistlberger, G., S. M. Borisov, and I. Klimant. 2009. Enhancing performance in optical sensing with magnetic nanoparticles. *Sens. Actuator. B* 139: 174–180.

Mock, J. J., D. R. Smith, and S. Schultz. 2003. Local refractive index dependence of plasmon resonance spectra from individual nanoparticles. *Nano Lett.* 3: 485–491.

Mohanpuria, P., N. K. Rana, and S. K. Yadav. 2008. Biosynthesis of nanoparticles: Technological concepts and future applications. *J. Nanopart. Res.* 10: 507–517.

Ohring, M. and S. P. Baker. 2002. *Materials Science of Thin Films*. (2nd ed.). Cambridge: Academic Press.

Orendoff, C. J., L. Gearheart, N. R. Jana, and C. J. Murphy. 2006. Aspect ratio dependence on surface enhanced Raman scattering using silver and gold nanorod substrates. *Phys. Chem. Chem. Phys.* 8: 165–170.

Polte, J. 2015. Fundamental growth principles of colloidal metal nanoparticles—A new perspective. *Cryst. Eng. Comm.* 17: 6809–6830.

Pong, B. K., H. I. Elim, J. X. Chong, W. Li, B. L. Trout, and J. Y. Lee. 2007. New insights on the nanoparticle growth mechanism in the citrate reduction of gold(III) salt: Formation of the Au nanowire intermediate, and its nonlinear optical properties. *J. Phys. Chem. C* 111: 6281–6287.

Rycenga, M., X. Xia, C. H. Moran et al. 2011. Generation of hot spots with silver nanocubes for single-molecule detection by surface-enhanced Raman scattering. *Angew. Chem. Int. Ed.* 123: 5587–5591.

Sannomiya, T. and J. Vörös. 2011. Single plasmonic nanoparticles for biosensing. *Trend. Biotechnol.* 29: 343–351.

Sawoo, S., D. Srimani, P. Dutta, R. Lahiri, and A. Sarkar. 2009. Size controlled synthesis of Pd nanoparticles in water, and their catalytic application in C-C coupling reactions. *Tetrahedron* 65: 4367–4374.

Schlüter, M., T. Hentzel, C. Suarez et al. 2014. Synthesis of novel palladium (0) nanocatalysts by microorganisms from heavy-metal-influenced high-alpine sites for dehalogenation of polychlorinated dioxins. *Chemosphere* 117: 462–470.

Self, C. H. 1985. Enzyme amplification—A general method applied to provide an immunoassisted assay for placental alkaline phosphatase. *J. Immunol. Method.* 76: 389–393.

Shahverdi, A. R., S. Minaeian, H. R. Shahverdi, H. Jamalifar, and A. A. Nohi. 2007. Rapid synthesis of silver nanoparticles using culture supernatants of *Enterobacteria*: A novel biological approach. *Process Biochem.* 42: 919–923.

Sharma, A. K. and B. D. Gupta. 2007. Metal-semiconductor nanocomposite layer based optical fiber surface plasmon resonance sensor. *J. Opt. A-Pure Appl. Op.* 9: 180.

Sharma, S., S. P. Usha, A. M. Shrivastav, and B. D. Gupta. 2017. A novel method of SPR based SnO_2: GNP nano-hybrid decorated optical fiber platform for hexachlorobenzene sensing. *Sens. Actuator. B* 246: 927–936.

Shrivastav, A. M., S. P. Usha, and B. D. Gupta. 2016. A localized and propagating SPR, and molecular imprinting based fiber-optic ascorbic acid sensor using an in situ polymerized polyaniline-Ag nanocomposite. *Nanotechnology* 27: 345501.

Siegel, J., O. Kvitek, P. Ulbrich, Z. Kolska, P. Slepicka, and V. Svorcík. 2012. Progressive approach for metal nanoparticle synthesis. *Mater. Lett.* 89: 47–50.

Song, J. Y., E. Y. Kwon, and B. S. Kim. 2010. Biological synthesis of platinum nanoparticles using *Diopyros kaki* leaf extract. *Bioprocess Biosyst. Eng.* 33: 159–164.

Sriram, M., K. Zong, S. R. C. Vivekchand, and J. J. Gooding. 2015. Single nanoparticle plasmonic sensors. *Sensors* 15: 25774–25792.

Srivastava, S. K., V. Arora, S. Sapra, and B. D. Gupta. 2012. Localized surface plasmon resonance based fiber optic U-shaped biosensor for the detection of blood glucose. *Plasmonics* 7: 261–268.

Stewart, M. E., C. R. Anderton, L. B. Thompson et al. 2008. Nanostructured plasmonic sensors. *Chem. Rev.* 108: 494–521.

Stiles, P. L., J. A. Dieringer, N. C. Shah, and R. P. Van Duyne. 2008. Surface-enhanced Raman spectroscopy. *Annu. Rev. Anal. Chem.* 1: 601–626.

Sun, Y. and Y. Xia. 2002a. Shape-controlled synthesis of gold and silver nanoparticles. *Science* 298: 2176–2179.

Sun, Y. and Y. Xia. 2002b. Increased sensitivity of surface plasmon resonance of gold nanoshells compared to that of gold solid colloids in response to environmental changes. *Anal. Chem.* 74: 5297–5305.

Suresh, A. K., D. A. Pelletier, W. Wang et al. 2010. Silver nanocrystallites: Biofabrication using *Shewanella oneidensis*, and an evaluation of their comparative toxicity on gram-negative, and gram-positive bacteria. *Environ. Sci. Technol.* 13: 5210–5215.

Tabassum, R. and B. D. Gupta. 2015a. Surface plasmon resonance based fiber optic hydrogen gas sensor utilizing palladium supported zinc oxide multilayers, and their nanocomposite. *App. Opt.* 54: 1032–1040.

Tabassum, R. and B. D. Gupta. 2015b. Fiber optic manganese ions sensor using SPR and nanocomposite of ZnO–polypyrrole. *Sens. Actuator. B* 220: 903–909.

Tabassum, R. and B. D. Gupta. 2016. Simultaneous estimation of vitamin K_1 and heparin with low limit of detection using cascaded channels fiber optic surface plasmon resonance. *Biosens. Bioelectron.* 86: 48–55.

Tabassum, R., P. Kaur, and B. D. Gupta. 2016. Tuning the field distribution and fabrication of Al@ZnO core shell nanostructure for SPR based fiber optic phenyl hydrazine sensor. *Nanotechnology* 27: 215501.

Tadanaga, K., K. Morita, K. Mori, and M. Tatsumisago. 2013. Synthesis of monodispersed silica nanoparticles with high concentration by the Stöber process. *J. Sol-Gel Sci. Technol.* 68: 341–345.

Tseng, K. H., J. C. Huang, C. Y. Liao, D. C. Tien, and T. T. Tsung. 2009. Preparation of gold ethanol colloid by the arc discharge method. *J. Alloy Compd.* 472: 446–450.

Tsuji, T., K. Iryo, N. Watanabe, and M. Tsuji. 2002. Preparation of silver nanoparticles by laser ablation in solution: Influence of laser wavelength on particle size. *Appl. Surf. Sci.* 202: 80–85.

Tsuji, T., T. Kakita, and M. Tsuji. 2003. Preparation of nanosize particle of silver with femtosecond laser ablation in water. *Appl. Surf. Sci.* 206: 314–320.

Turkevich, J., P. C. Stevenson, and J. Hillier. 1951. A study of the nucleation, and growth processes in the synthesis of colloidal gold. *Discuss. Faraday. Soc.* 11: 55–75.

Usha, S. P., A. M. Shrivastav, and B. D. Gupta. 2017. Silver nanoparticle noduled ZnO nanowedge fetched novel FO-LMR based H_2O_2 biosensor: A twin regime sensor for in-vivo applications and H_2O_2 generation analysis from polyphenolic daily devouring beverages. *Sens. Actuator. B* 241: 129–145.

Usha, S. P., A. M. Shrivstav, and B. D. Gupta. 2016. FO-SPR based dextrose sensor using Ag/ZnO nanorods/GO_x for insulinoma detection. *Biosens. Bioelectron.* 85: 986–995.

Vilchis-Nestor, A. R., V. Sánchez-Mendieta, M. A. Camacho-López, R. M. Gómez-Espinosa, M. A. Camacho-López, and J. A. Arenas-Alatorre. 2008. Solventless synthesis, and optical properties of Au, and Ag nanoparticles using *Camellia sinensis* extract. *Mater. Lett.* 62: 3103–3105.

Wiley, B. J., S. H. Im, Z. Y. Li, J. McLellan, A. Siekkinen, and Y. Xia. 2006. Maneuvering the SPR of silver nanostructures through shape-controlled synthesis. *J. Phys. Chem. B* 110: 15666–15675.

Xu, J., G. A. Cheng, and R. T. Zheng. 2010. Controllable synthesis of highly ordered Ag nanorod arrays by chemical deposition method. *Appl. Surf. Sci.* 256: 5006–5010.
Zhang, D., L. Qi, J. Yang, J. Ma, H. Cheng, and L. Huang. 2004. Wet chemical synthesis of silver nanowire thin films at ambient temperature. *Chem. Mater.* 16: 872–876.
Zhang, J., J. Du, B. Han, Z. Liu, T. Jiang, and Z. Zhang. 2006. Sonochemical formation of single-crystalline gold nanobelts. *Angew. Chem.* 118: 1134–1137.
Zou, J., Y. Xu, B. Hou, D. Wu, and Y. Sun. 2007. Controlled growth of silver nanoparticles in a hydrothermal process. *China Particuology* 5: 206–212.

5

Semiconductor Metal Oxide Sensors

Materials can be classified into conductors, semiconductors, and insulators on the basis of band theory of solids. For bulk materials, the energy states are defined in the forms of molecular orbitals/bands instead of discrete energy levels. The band having molecular orbitals filled/partially filled is called valence band, while the one having empty molecular orbitals is called conduction band. The energy of conduction band is higher than that of the valence band and it allows electrons from valence band to flow to it. The energy gap/bandgap between the valence and conduction bands decides whether the material is a conductor/semiconductor/insulator. In the case of conductors, the bandgap is 0 (e.g., metals), but for insulators it is >3 eV. Semiconductors are materials having their bandgap energy between 0 and 3 eV. The energy band diagrams of conductor, semiconductor, and insulator are shown in Figure 5.1.

Semiconductors hold a better capability of varying the electronic concentration in their conduction band. Semiconductor materials can be elemental, carbon/organic form, binary compounds, or oxides. Some elements of periodic table such as silicon (Si), germanium (Ge), selenium (Se), sulfur (S), phosphorous (P), and tellurium (Te) mainly belong to group IV of periodic table and come under the category of elemental semiconductors. These are the elements in which each atom is surrounded by two (e.g., S), three (e.g., P), or four (e.g., Si, Ge) nearest atoms to give twofold, threefold, or fourfold coordinates. Carbon in the form of fullerene or nanotube shows semiconducting properties. Binary compound semiconductors are formed by grouping the elements from groups III and V, II and VI, I and VII, or IV and VI, to behave like group IV elements. Due to the transfer of electrons from one group to another in binary compounds, the bond becomes ionic in nature which affects the semiconductor behavior (Peter and Cardona 2010). Oxides, in general, are insulators but some combinations of metals/transition metals and oxygen in the periodic table provide metal oxides such as Cu_2O, ZnO, and Fe_2O_3 which are conventional semiconductors. These semiconductor metal oxides (SMOs) have a number of properties which suit in the sensing of chemical and biological analytes. Therefore, in this chapter, we shall focus on the sensors utilizing SMOs.

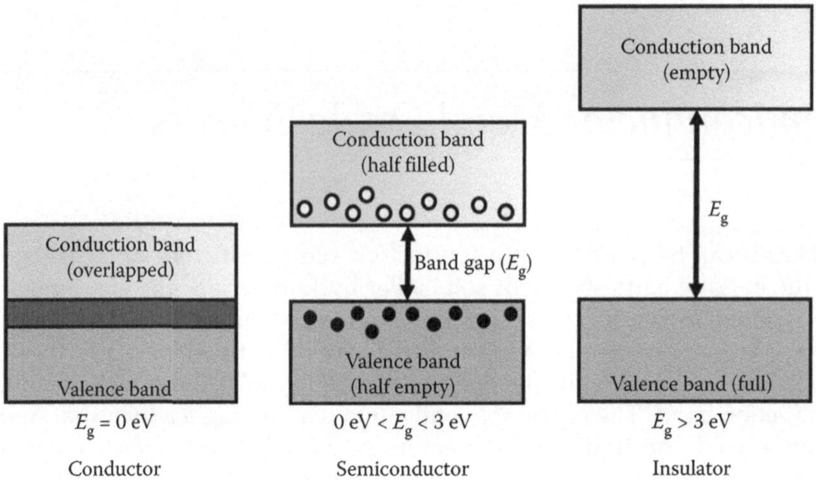

FIGURE 5.1
Energy band diagram of conductor, semiconductor, and insulator.

5.1 Role of SMO in Sensor Applications

The amplest materials in the Earth's crust are metal oxides. Their general structure has either oxygen surrounding the metal ions or metal ion surrounding the oxygen, where the force of bond holds the SM^+ and O^{2-} together (Henrich and Cox 1996). SMOs have driven the interest of researchers and industrialists due to their wide range of advantages in comparison to other semiconductors of elemental and organic forms. The disadvantages of elemental semiconductors include limited optical transparency and conductivity, poor charge carrier mobility, and mechanical flexibility. These shortcomings are overcome in SMOs with their high-optical transparency and conductivity (which sounds contradictory), distinctive charge carrier mobility, proper stoichiometry (ratio of various kinds of atoms present), excellent mechanical flexibility, pressure tolerance, and compatibility with inorganic and organic materials (Yu et al. 2016). SMOs, due to these attractive properties, are used in optical and optoelectronic applications in thin film transistors, solar cells, heat-reflecting films, protective layers, and sensors. Optical transparency and conducting properties of metal oxides make them hold an efficient role in practices in both visible and infrared range of electromagnetic spectrum. The spectral discerning capability of metal oxides is useful for diverse implementations. An example of this is the usage of metal oxide film as heat-reflective coating. Metal oxide films are also used as heat insulation coatings with unrestrained light passage. This is because of their property of complete light transmission in the visible range, where they behave like insulators and the reflection of light in the infrared (heat radiation) range, where they behave as metals.

One of the principal applications of SMOs is sensing. Chemical sensing, especially gas, and biosensing have been reported in the literature using SMOs. For chemical sensors, SMOs can be employed as a transducer element and/or as a recognition element, while in the case of biosensors, they mainly play the role of a transducer element and/or as a matrix layer for biorecognition element immobilization. In this chapter, we shall discuss both in detail. The initial sections are devoted to the SMO-based gas sensors, since the major application area of SMOs in the field of sensing is environmental monitoring. SMOs facilitate the direct adsorption of gas molecules on their surface. The adsorption changes the conductivity/conduction band charge carrier concentration of SMO due to the exchange of charge carriers with the adsorbed gas species. At the same time, the interaction can be made reversible as well. This makes SMO a suitable candidate for gas sensing. The demand needed to be met by the SMO for gas-sensing purpose is that its structure should be porous enough to support the maximum interaction of gas and the properties should be controlled by surface and subsurface (Korotcenkov 2007). The properties can be physical, electrical, mechanical/structural, or optical that measure the variation due to the interaction of gas on the surface and subsurface states. Hence, gas sensors have been reported in literature using different designs depending on the property on the basis of which the sensor's transducer element is chosen. Further, depending on the various requirements such as the gas to be sensed, its location and purpose, and the performance of the sensor to be met like selectivity and reliability, the SMO material and its structure are chosen. Hence, the choice and design of SMO material in sensor realization require more research and development.

Optical gas sensors designed using SMO work on the principle of conductometric detection. The interaction of gas molecules adsorbed on the SMO surface changes its conductance/resistance which is measured by the optical transducer. The method offers excellent sensitivity, compatibility with portable devices, cost-effectiveness, low-response time, stability, and reliability in comparison to electrochemical, thermal conductive, and catalytic combustion-based gas sensors (Korotcenkov 2007). The conductometric gas-sensing principle is based on the interaction of gas molecules with the SMO surface states, subsurface states, and defects in the metal oxide structure. The partially filled energy states found in the gap between the valence band and conduction band of a semiconductor are called surface states. Defect is the irregularity in the crystal structure/deviation from its ideal crystal structure. Defects are usual on SMO surfaces/subsurfaces than in bulk of the material. These depend on the fabrication process, temperature conditions, interface distances, and grain (crystallite) boundaries of the fabricated SMO (Korotcenkov 2005). Certain SMOs possess a large number of defects in the bulk of the material which cause chemical and catalytic reactions with the gas molecules. Bulk defects can be point defects or extended defects. Point defects include vacancies or interstitials in lattice and dopants or impurities or adsorbed oxygen species O^{2-} that occupy the interstitial sites. For example, chemisorption, defect in electronic structure, oxygen vacancies, corrosion, etc. come under point defects. Dislocations and planar defects (grain boundaries, surface) form extended type of crystal defects (Yu et al. 2016). Point defect in ZnO makes it an important material for gas-sensing applications. The defects in SMOs usually occur in the form of pairs or combinations to ensure the electrical neutrality of the crystal. A combination of vacancy and interstitial defect is called the Frenkel defect, while a pair of vacancies or interstitials is called the Schottky defect.

The interaction of the gas molecule with SMO can be because of one or more of the possible interactions listed below:

1. Redox reaction
2. Direct adsorption
3. Adsorption and reaction with the surface states, chemical reaction with the oxygen species, or point defects
4. Catalytic interaction due to the presence of dopants or water vapor

These interactions vary the conductivity of SMO. As we know, the conductivity of a semiconductor is related to its dielectric constant/permittivity through the relation

$$\varepsilon_{SMO} = \varepsilon_r + i\varepsilon_i = \varepsilon_r + i\frac{\sigma}{\omega\varepsilon_0} \tag{5.1}$$

where ε_{SMO} is the material permittivity with real part ε_r and imaginary part ε_i, σ is the material conductivity, ε_0 is the material permittivity in vacuum, and ω is the frequency. Thus, a change in the conductivity changes the permittivity which is the square of refractive index expressed by

$$\varepsilon_{SMO} = (n_{SMO})^2 = (n + ik)^2 \tag{5.2}$$

Hence, the real (n) and imaginary (k) parts of the refractive index change with the change in the permittivity which can be detected efficiently by the optical transducer using the techniques of surface plasmon resonance (SPR) and lossy mode resonance (LMR).

Depending on the property and reactivity of the gas molecule and the SMO, the sensor can be designed. The property of the semiconductor to adsorb oxygen species on its surface soon after its preparation plays a major role in the conductometric detection of gases. The adsorption of oxygen can be physisorption or chemisorption. The adsorption of electrically neutral species or the adsorption without any ion exchange between gas and SMO is called physisorption, while the adsorption of charged oxygen species that results in ion exchange is called chemisorption (Bielański and Najbar 1972). The physisorption does not involve any change in the surface potential of SMO, whereas chemisorption involves the change in the SMO surface potential (Korotcenkov 2007).

The adsorption mechanism is similar for analyte gases as well, that can diffuse through the SMO structure. Some gases may result in physisorption, while others may result in chemisorption. A distinct combination of the mechanism of adsorption and desorption for the oxygen and gas species should be ensured by the SMO material chosen for sensing. Figure 5.2 shows the schematic illustration of adsorption/desorption on SMO surface with diffusion of gas in SMO which affects the gas sensor response.

The chemical and catalytic reaction between the analyte gas and SMO is another reason for the change in surface conductivity. The reaction takes place not only with the surface oxygen or adsorbed oxygen but also with the oxygen defects in the interstitial sites of SMO (Korotcenkov 2007). This happens due to the property of the gas to diffuse through the structure. The diffusion may take place through the bulk surface, space between grains, inside and between the agglomerates of grains as shown in Figure 5.2. The choice of SMO to support such an interaction with the sensing analyte makes the sensor more sensitive and selective in comparison to simple adsorption-based method of sensing. Thus, there are various SMO properties that regulate the

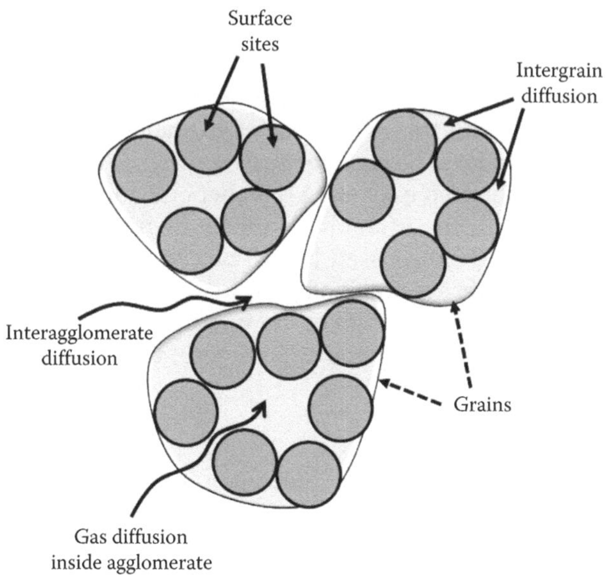

FIGURE 5.2
Schematic illustration of adsorption/desorption and diffusion kinetics of gas sensing using SMO.

interaction of material with the analyte gas and hence the sensor's performance, which will be discussed in the following section.

5.2 Properties Supporting Sensing

The transmission and reflection properties of SMOs decide their role in the optical sensor depending on the sensor's spectral region of operation (Dawar and Joshi 1984). The transmissivity and reflectivity of an SMO depends on the concentration of carriers present in the material. An increase in the carrier concentration increases the reflectivity and decreases the transmission of light through the material.

SMOs can be used to play dual role in both visible and infrared regions of sensing operation. In the visible regime, they perform as LMR-transducer element as well as a recognition element, while in the infrared regime they are used as an SPR-transducer and -recognition element. Since in both regimes SMOs are used as the transducer and sensing layer, the sensor fabrication is much easier compared to other material-based techniques. The interaction of SMO with the gas molecules changes the SMO's optical property (refractive index) which changes the condition to satisfy the resonance. By analyzing the resonance condition, the sensor is calibrated. Apart from the dual role of operation of SMO, it can also be utilized for gas sensing in two different ways in the visible region. One way can be realized by utilizing SMO as both recognition and transducer element. The other way is by combining the SMO–gas-recognition layer with a plasmonic transducer layer. In both the cases, SMO interacts with the gas molecules to change its conductivity and hence the refractive index. The change in refractive index is measured by the plasmonic metal/SMO itself exploiting the phenomenon of SPR/LMR.

The SMO's material properties that support, regulate, and evaluate the sensing performance are discussed below. Since the chapter mainly deals with optical sensing of gases and biological analytes using SMOs, the properties are discussed from that point of view. All the properties exhibited by SMO depend on its fabrication/deposition process and the substrate used.

5.2.1 Surface and Structure

The exposed/active surface for gas interaction, grain size of the deposited bulk/nanostructure, and the surface area to volume ratio of SMO are the primary parameters for the efficient sensing of gases by SMO (Korotcenkov et al. 2005). By optimizing the bulk and nanostructure size, shape, porosity, and orientation of crystal planes, sensor's performance can be maximized. Reports in the literature have classified the surface and structural parameters controlling the gas-sensing properties into two main groups: SMO geometry and SMO chemical composition (Korotcenkov 2005). Geometry parameters include grain size, distance between grains, thickness, porosity, agglomeration, area of inter-grain boundary, and distances between inter-grain boundaries. The chemical composition considers elemental and phase composition, conductivity, surface design, dopants, and stoichiometry of components of SMO. An agglomeration in the structure, mainly in the case of SMO nanomaterials, limits the permeability of gas and hence the interaction. The film thickness and agglomeration controls the response time as well as the recovery of the sensor. For a fairly thin layer of bulk/nanostructure SMO-based gas sensor with minimal agglomeration, the rate of response will be high (Korotcenkov et al. 2003). If the grain size of the deposited layer is very small and they are agglomerated, then the interaction of analyte gas with the layer will become minimum due to less permeability of gas through layer (Brinzari et al. 2001). The addition of dopants to change the composition changes the appearance, grain size, concentration of charge carriers, and physical, chemical, and surface properties of SMO. The ratio of dopant with SMO should be optimized according to the analyte gas and its reaction with the effective composition of SMO.

Considering all of the above said surface and structural parameters, the optimizations of parameters are required to bring out the best of a sensor. However, all the sensor parameters cannot be optimized simultaneously. A change in one parameter concurrently changes the other parameter's effect. Hence, depending on the application or gas to be sensed, the parameter that should be given much importance needs to be decided in the first step. This, in turn, depends on the demands to be met in a sensor such as high selectivity, sensitivity, short response time, etc. For conductometric-based gas sensors, the parameter of choice depends on the oxidizing/reducing property of the gas with the SMO and the interactive behavior which decides adsorption–desorption/redox reaction/chemical, or catalytic actions.

The SMOs can be classified into transitional and nontransitional type based on their electronic structure. Fe_2O_3 and Cr_2O_3 are the examples of transitional type, while ZnO, SnO_2, and Al_2O_3 are the examples of nontransitional type. In nontransitional type, again the SMOs are classified into pretransitional like Al_2O_3 and posttransitional like ZnO and SnO_2. Transitional SMOs have limited application in gas sensing due to less stability and conductivity reasons. Creating defects in nontransitional SMOs are reported to be easier, which makes their role important in gas sensing (Korotcenkov 2007). In the case of pretransition metal oxides, the bandgap is much larger. Hence, it is difficult to create free electrons and holes, which make them almost inert in reactions. For posttransitional SMOs, the valance orbital is d-shell with a configuration of d^{10} and bandgap lies in the range 3–4 eV. Due to the feasibility in changing this electronic configuration, these SMOs actively participate in redox reactions. Posttransitional SMOs easily get reduced by their interaction with reducing gases, resulting in the generation of free charge carriers. The increase in charge carrier concentration increases the conductivity. The reaction of such SMOs with oxidizing gases occurs only when the SMO surface is made lacking/insufficient of oxygen. An example of this is the choice of ZnO for the detection of reducing gas such as H_2S (Usha et al. 2015a).

5.2.2 Conductivity

In addition to surface interaction and matrix parameters, there are many electrophysical and chemical parameters which monitor the SMO sensing operation. Conductivity is one among such parameters. As we know, semiconductors can be n-type or p-type. In an n-type semiconductor, the majority charge carriers are electrons, while in p-type holes are the majority charge carriers. Examples of common n-type SMOs are ZnO, SnO_2, and In_2O_3, while p-type SMOs are NiO and CuO. The interacting gases can be reducing (electron donors) or oxidizing (electron acceptors) types. Reducing gases include H_2S, CO, and H_2, while the gases like O_2, Cl_2, and NO_2 are oxidizing. The interaction of n-type SMO with a reducing gas increases the number of electrons in SMO and hence the conductivity increases, while its interaction with an oxidizing gas reduces the conductivity. In the case of p-type SMOs, an increase in the conductivity occurs for an interaction with oxidizing gas and vice-versa (Fine et al. 2010). Generally, n-type SMOs are employed more for the sensing of gases in comparison to p-type SMOs. p-Type SMOs have some limitations such as lower stability due to susceptibility to lose interstitial oxygen into atmosphere, less thermal reliability, poor reproducibility, and decrement in surface conductivity while interacting with reducing gases which make the material noncompatible to use for sensor applications.

The concentration of defects needs to be limited for a material to be used in a conductometric sensor. This is because when the concentration of such defects is higher, the material becomes highly conductive (similar to metals) in nature and the probability of the material surface to interact with the analyte gas becomes less. This affects its sensing performance adversely. The ideal range of defects in SMOs is 10^{17}–10^{20} cm^{-3} (Korotcenkov 2007). SMOs having free charge carrier concentration $<10^{16}$ cm^{-3} do not support conductometric sensing.

5.2.3 Catalytic/Chemical Activity and Stability

In many gas sensors, catalytic or chemical reaction of the analyte gas takes place with the SMO surface and adsorbed oxygen. These reactions change the concentration of point defects because of the involvement of interstitial oxygen defects of the polycrystalline or nanocrystalline SMOs. This, in turn, changes the conductivity of SMOs. The chemical interaction of H_2S gas with ZnO and catalyst reaction of palladium-doped tin oxide with CH_4 gas are some of the examples (Dutronc et al. 1993, Usha et al. 2015a). The advantage of such reaction-based gas sensor has been the high selectivity for the analyte gas since only the preferred gas could undergo the desired kinetics. Considering environmental safety, the low/atmospheric temperature kinetics are advised for such reaction-based gas sensing. This is because, for industrial applications, detection of flammable and toxic gases at atmospheric conditions is required. Catalytic activity-based sensors are usually designed with SMOs such as MnO_2 and NiO. The materials such as ZnO and Cu_2O are highly active and most promising SMOs for gas sensing as discussed above and therefore do not require any catalyst to support the sensing performance (Korotcenkov 2007). Thus, catalytic activity in gas sensing is a supporting parameter and not the response determining parameter in all the cases. The sensor and sensing material used, especially for biomedical applications, should have extreme chemical stability even at high temperatures. It should not be reactive with any other blood elements/ human body. Metal oxide semiconductors are more stable in comparison to other semiconductors and metals, which make them a better candidate in the industrial applications.

5.2.4 Sensitivity and Reversibility

The sensitivity of sensor depends on the surface properties, chemical, and catalytic reactions as discussed above. The designed sensor should be highly sensitive to the analyte gas with respect to the possible interfering gases. The effect of water vapor on SMO surface is a considerable issue affecting the sensor performance. Sometimes sensor needs to work on conditions of higher humidity. Thus, depending on the place and application in which sensor is supposed to work, the sensor's ability to resist water vapor adsorption should be evaluated. It has been reported that there is a competitive adsorption between oxygen and water species on certain SMO surface. Hence, such SMO-based sensors behave divergently in dry and humid conditions. An example is the competitive adsorption between hydroxyl group and other gases like methane and oxygen on the tin oxide surface hindering the gas adsorption. Hence, for better practical application in humid atmosphere, the sensor should be designed insensitive to water vapor (Korotcenkov 2005).

In addition to sensitivity, reversibility/recovery is also a desired property for a sensor to be commercialized. SMO sensors can be easily recovered from the adsorption and redox reaction occurring on their surface since the bonding in the by-products formed is generally weak in nature. The time that sensor needs to recover to its earlier stage is called recovery time. In addition to the response time of sensor, which is the time taken by the sensor to give a saturated output, recovery time is equally important. The sensor should possess easy recovery process and an acceptable time of recovery for it to be implemented in atmospheric monitoring applications especially in chemical industries, oil and gas companies, mining operations, etc.

5.3 Nanostructured Metal Oxides

Like nanostructures of any material, metal oxide nanostructures also possess unique physical, chemical, electrical, and electronic properties that differ from their bulk structure. These properties have been utilized to build novel sensors. For the desired analyte, the sensor can be designed to give enhanced performance by tuning the size, shape, and structure of the concerned SMO nanomaterial. The common metal oxide nanostructures employed for environmental monitoring and biomedical applications are that of iron oxide (Kaushik et al. 2008), zinc oxide (Usha et al. 2015a), tin oxide (Jia et al. 2005), titanium oxide (Liu et al. 2006), zirconium oxide (Liu et al. 2004), magnesium oxide (Xu et al. 2004), nickel oxide (Salimi et al. 2007), and indium tin oxide (Singh and Gupta 2010). The 1D nanostructures of SMOs have attracted much attention in sensing applications due to higher aspect ratio (surface area to volume ratio) and excellent chemical and thermodynamic stability at various conditions of operations (Arafat et al. 2012). By modifying the surface states, defects, and properties of the bulk material in the 1D nanostructures, an improved response of the sensor can be achieved. The modifications can be achieved by depositing the nanostructures using various methods, combining nanostructure with bulk layer, and reorganizing the nanostructure surface with other material coatings or dopants.

The preparation and deposition of the nanostructures result in the deformation in the crystalline structure of the material. These deformations deflect as discontinuity in the crystal lattice or symmetry of the crystalline structure which results in the defect formations on the boundary and interfaces of the grain structure in crystal. Such defects mean localized states for electrons and holes in the crystal. In nanostructures, these defects lead to variation in the material's electronic structure which changes the transport of electrons in them compared to bulk form (Anta 2012). The SMO parameters that influence the process of electron transport include surface disorder, structure of crystal, diameter of nanostructure, chemical configuration, and orientation of crystal axis. The conductivity of structure, in turn, depends on all these parameters. Similar to bulk structure, nanostructured SMOs also possess biocompatibility, stability, nontoxicity, high-electron transport, and conductivity that make them ideal as transducer layer, reaction mediator, immobilization matrix, or recognition element.

The optical, electrical, chemical, and structural properties of nanostructures depend on the conditions of their synthesis. The main synthesis methods for nanostructures are physical and chemical. In the physical and chemical methods of preparation, the processing route can be in various states of material such as solid, liquid, vapor, molten, and hybrid of these. Table 5.1 shows the physical and chemical methods of nanostructure preparation using different processing routes. Most of these methods, their advantages and limitations, have already been discussed in Chapter 4. The

TABLE 5.1

Common Nanostructure Synthesis Methods

	Method of Synthesis	
Processing Route	**Physical**	**Chemical**
Solid state	Thermal evaporation Carbothermal reduction Plasma etching Mechanical milling	Direct oxidation Solgel Chemical reaction
Liquid state	Electrospinning Laser ablation	Hydrothermal Drop coating Screen printing Electrochemical deposition Sonochemical Photo-catalytic reduction Microemulsion Hydrothermal with photocatalytic reduction
Gaseous state	Thermal evaporation RF sputtering Vapor-phase transport Aerosol Spray pyrolysis Pulse laser deposition	Molecular beam epitaxy Chemical vapor deposition
Molten state		Molten salt-based chemical solution
Hybrid state	Pulse laser deposition with electrospinning	Chemical vapor deposition with electrospinning

effects of these methods on the structure and properties of SMO nanostructures fabricated on optical substrate are different. The mostly used commercial method for SMO sensor fabrication is screen printing. In this method, a porous grid is placed over the substrate. The SMO nanomaterial to be printed, prepared in the form of a liquid, is pushed over the grid to deposit it. After deposition, the arrangement is heated to print the material over the required place and after this the grid is removed. For fast growth of nanostructure, chemical vapor deposition is used. This method provides regulation over porousness, film thickness, and interfacial and grain expanse properties of the SMO material. The methods of chemical vapor deposition and hydrothermal provide a uniform thin layer of SMO material over the substrate with good adhering property. However, hydrothermal method prepares porous-sparse structure that allows higher interaction with analyte in comparison to the dense layer formation using chemical vapor deposition method. Moreover, these methods such as vapor deposition, sputtering, and laser ablation are expensive (Fine et al. 2010).

For fabricating 1D nanostructures, the generally used method is hydrothermal. This method offers advantages of simple nano-growth mechanism with favorable growing conditions such as low temperature, growing period, and easier deposition on substrate (Arafat et al. 2012). The production of nanostructures is higher in liquid processing route compared to other routes. In spray pyrolysis, the chemical reactants for preparing SMO material are sprayed on to the substrate. The substrate is then heated so that the chemicals undergo reaction to form a layer over it (Fine et al. 2010). Solgel process is another one used for the synthesis which includes the preparation of solution in the first stage and gelation in the second stage. The solution consists of an insoluble metal oxide particles suspended in a suitable solvent. Gelation process involves the cross-linking of these metal oxide particles which changes their properties. One of the limitations of this method is the time needed to set the solution, which is usually performed in several steps. For biocompatible and environmental friendly applications, generally, liquid processing route-based chemical methods are used.

5.4 Mechanism of Gas Sensing and Applications

The mechanism of gas sensing using SMO materials is mainly due to chemisorption or chemical/catalytic reactions as discussed before. It is practically impossible to fabricate a stoichiometric (defect-free) SMO surface. These defects decide the chemisorption and reaction mechanism of gas sensing at differing pressure level. The chemisorption reactions are more likely to happen at low-atmospheric pressures, while catalytic reactions occur at high-atmospheric pressures. At high-atmospheric pressure, the number of reactions happening per unit space per unit time will be more that enhance the catalytic activities, which is assured by chemisorption mechanism at lower pressures (Henrich and Cox 1996).

In SMOs, chemisorption of oxygen takes place as soon as they have been prepared. The adsorption can occur in the form of molecular oxygen or charged oxygen species (Bielański and Najbar 1972, Usha et al. 2015a). This results in the production of negatively charged oxygen species with the emergence of positive-charged ions, neutralizing the crystal. The target gas molecules can be donor type or acceptor type. The acceptor types interact with the negatively charged oxygen species and defects which are basic in nature, while the donor types interact with the positive-charged ions (Henrich and Cox 1996).

The molecular oxygen adsorption on SMO surface does not involve charge transfer with the SMO. The adsorption of oxygen as charged species may occur in three forms such as superoxide (O_2^-), ion radical ($2O^-$), and peroxide ($2O^{2-}$). The formation of O^{2-} has been considered as the oxygen species that occupies the SMO crystal lattice (Bielański and Najbar 1972). These three oxygen species also generate free holes in order—1 hole, 2 holes, and 4 holes, respectively—to result in the production of positive-charged ions of

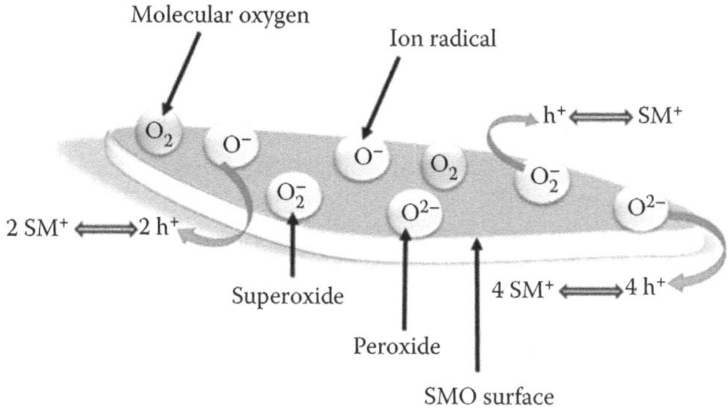

FIGURE 5.3
Chemisorption of oxygen on SMO surface.

SMO, here onward denoted as SM^+. The chemisorption of superoxide results in the formation of SM^+ with the release of a single hole. For ion radical and peroxide species, their chemisorption results in the formation of $2SM^+$ and $4SM^+$ due to the release of 2 holes and 4 holes, respectively as shown in Figure 5.3. These chemisorbed oxygen species act as the active sites in addition to the defects and react with the target gas of acceptor type to create by-products, while positive-charged species interacts with target gas of donor type. The by-products formed can be easily withdrawn from SMO surface, which makes SMO and these interactions a suitable candidate for gas-sensing applications.

The chemical reactions such as reduction and oxidation occur on SMO surface when gas molecules interact with SMO material. Such reactions involve charge transfer, which means the addition or removal of oxygen from interstitial sites/defects. Thus, defects support chemical redox reactions in SMOs. The condition of such reactions to happen in an SMO is defined in most of the work as the presence of vacancies/point defects and/or the adsorbed oxygen species (Henrich and Cox 1996). Catalytic reactions occur when SMO material mainly has dopants. Another possibility is the water vapor adsorption that may happen on SMO surface which acts as a catalyst. Thus, the chemisorption of water vapor on SMO surface modifies the properties of SMO and catalyzes the interaction between SMO and target gas (Korotcenkov 2007).

These interactions change the SMO conductivity which in turn modifies the dielectric constant/refractive index of the SMO material, as discussed in Section 5.1. The change in refractive index can be correlated to the measurand (concentration, pressure, temperature, etc.) in the mechanism of gas-sensing applications. The sensors based on such mechanism have been reported in the literature for environmental monitoring using the techniques of SPR, LSPR, and LMR. The theories of SPRs and LMR have already been explored in Chapter 2. In the case of SPR/LMR-based SMO gas sensors, the interaction of gas molecules on SMO material may either increase or decrease the refractive index of the material. The change will be transduced optically by techniques of SPR/LMR by the shift in resonance wavelength and the change in the depth of the resonance spectrum. In both SPR and LMR phenomena, the increment/decrement in spectral depth and right/left shift in resonance wavelength correspond to the increase/decrease in imaginary and real parts of the refractive index of the sensing material, respectively.

5.4.1 SPR-Based SMO Gas Sensors

In the case of SPR-based SMO gas sensors, SMO is used only as the sensing material if the operating wavelength of the sensor is in the visible region of the electromagnetic spectrum. For infrared region of operation, SMO material can act as both sensing material and transducing element, making use of the material's interactive and reflective (plasmonic or metal-like) properties, respectively. However, because of high cost of the instrumentation in infrared region of operation, the usage of SPR-SMO-based sensor in the visible region is encouraged. Such sensors are designed by coating a thin film of plasmonic material (for example, silver and gold), over the substrate followed by the coating of an overlayer of SMO material to interact with the target gas. The target gas changes the dielectric constant/refractive index of the SMO layer. Some of gas sensors reported in the literature using SPR technique implementing SMO as sensing element have been discussed in this section to give readers a brief idea on gas sensing using SMO material and SPR technique.

Using tin oxide (SnO_2) as sensing element, an SPR-based ammonia gas sensor designed on fiber-optic substrate was reported (Pathak et al. 2015). The sensor probe was fabricated by coating a silver film of 40 nm thickness over a 1 cm unclad portion in the middle of about 20-cm-long plastic-clad silica fiber. To sense the gas, a 15-nm-thick SnO_2 layer was coated over the silver-coated portion of the probe. The deposition of both the layers was performed using thermal evaporation technique. The gas-sensing setup and the arrangements used for characterizing the sensor probe are given in Chapter 4. The adsorption of oxygen species on SnO_2 surface and the lattice defects on it play considerable roles in the sensing of ammonia gas. The ammonia gas interacting with the SnO_2 surface initially undergoes oxidation in the presence of already adsorbed oxygen species resulting in the generation of NH_x. The generated NH_x species start diffusing into the structure to undergo further oxidation to produce N_2 and H_2O. The evolution of N_2 by releasing electrons back to the surface results in an increment in the conductivity of SMO. Additionally, these intermediate products N_2, H_2O, and NH_x start reacting with SnO_2 resulting in the formation of tin nitride and tin hydroxide. These intermediates and by-products increase the real and imaginary parts of the refractive index of the sensing medium which was depicted in the SPR spectra with a redshift in the resonance wavelength and increase

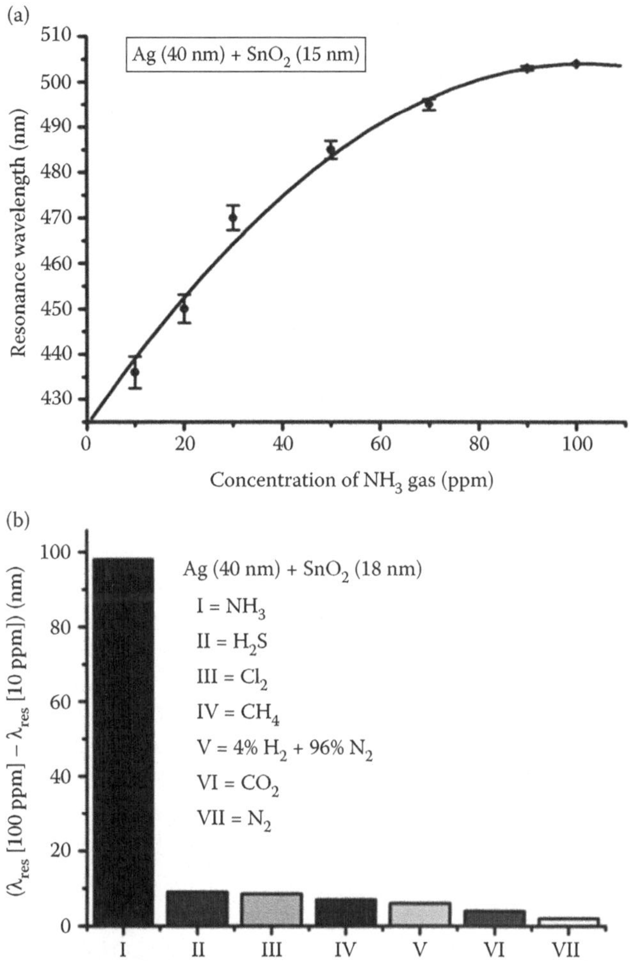

FIGURE 5.4
(a) The calibration curve showing resonance wavelength versus concentration extracted from SPR spectra obtained for the ammonia gas sensor and (b) selectivity test of the designed probe for gases within a concentration range from 10 to 100 ppm. (Reprinted from Pathak, A. et al. 2015. Fiber optic ammonia sensor using Ag/SnO_2 thin films: Optimization of thickness of SnO_2 film using electric field distribution and reaction factor. *App. Opt.* 54: 8712–8721. With permission from The Optical Society, OSA.)

in the depth of the SPR curve, respectively. Figure 5.4a shows the resonance wavelength extracted from each SPR spectrum obtained for the ammonia gas concentration varied in the probe's vicinity from 10 to 100 ppm. A maximum shift of 72 nm in the resonance wavelength was reported. The adsorption and chemical reactions associated with SnO_2 surface did not produce a permanent layer and hence the reactions were made easily reversible using the nitrogen purging.

The involvement of chemical reactions between the SnO_2 surface and ammonia gas and the interaction of ammonia gas with oxygen species on SnO_2 are selective to this particular gas. This was cross-checked by verifying the sensor probe's behavior in the presence of other oxidizing and reducing gases. Figure 5.4b shows the results of selectivity test conducted by experimenting probe in the presence of various other gases. A fair selectivity has been reported for ammonia gas due to the chemical reaction's support involved in the interaction of ammonia with SnO_2. The other gases were found to be reacting with the oxygen species only causing a minute change in the sensing medium refractive index and a small shift in the resonance wavelength for various gases with a concentration change from 10 to 100 ppm.

Sometimes the doping of one type of metal oxide in another type of metal oxide is used for the gas-sensing application. An example of such doping-based sensor is hydrogen sulfide (H_2S) gas sensor that uses indium tin oxide (ITO = In_2O_3: SnO_2) doped with nickel oxide (NiO) (Mishra et al. 2014). NiO possesses surplus oxygen which increases the oxygen defects/vacancies in ITO structure. An increased defect quantity increases the number of active sites for the interaction with the target gas. The sensor probe for H_2S gas was fabricated similar to ammonia gas sensor discussed above. The unclad portion of the fiber was coated with a 40-nm-thick silver film followed by the deposition of NiO-doped ITO film. A 7-nm-thick layer of NiO-doped ITO was coated having 10% of NiO and 90% of ITO. The sensing layer thickness and doping concentration were optimized to obtain the maximum sensitivity of the sensor. The gas-sensing setup for H_2S was the same as discussed in Chapter 4. The sensor was characterized for the concentration range of 10–100 ppm for which a redshift of around 104 nm was recorded. The interaction of H_2S gas with the sensing element results in the generation of sulfide layer given by the reaction:

$$[(10\%)\text{NiO} + (90\%)\text{ITO}] + \text{H}_2\text{S} \Leftrightarrow [(10\%)\text{NiO} + (90\%)\text{ITO}]\text{S} + \text{H}_2\text{O} \quad (5.3)$$

The by-products of the reaction, which is in the form of a complex, cover the surface of the sensing probe. This results in a change in the effective refractive index of the sensing layer which in turn varies the SPR spectral characteristics. Figure 5.5a and b shows the SPR spectra and calibration curve (resonance wavelength vs. H_2S concentration) of the sensor, respectively. With an increment in the H_2S gas concentration, an increased interaction occurs between NiO-doped ITO film and H_2S gas resulting in a thicker sulfide layer formation. The redshift of resonance wavelength and increase in the depth of the SPR spectrum confirmed the respective increments in real and imaginary parts of the effective refractive index of the sensing layer. An analysis of the calibration curve shows a decrement in the rate of change in resonance wavelength with the increment in H_2S gas concentration. A maximal reactive behavior was shown by the sensor probe for H_2S concentration

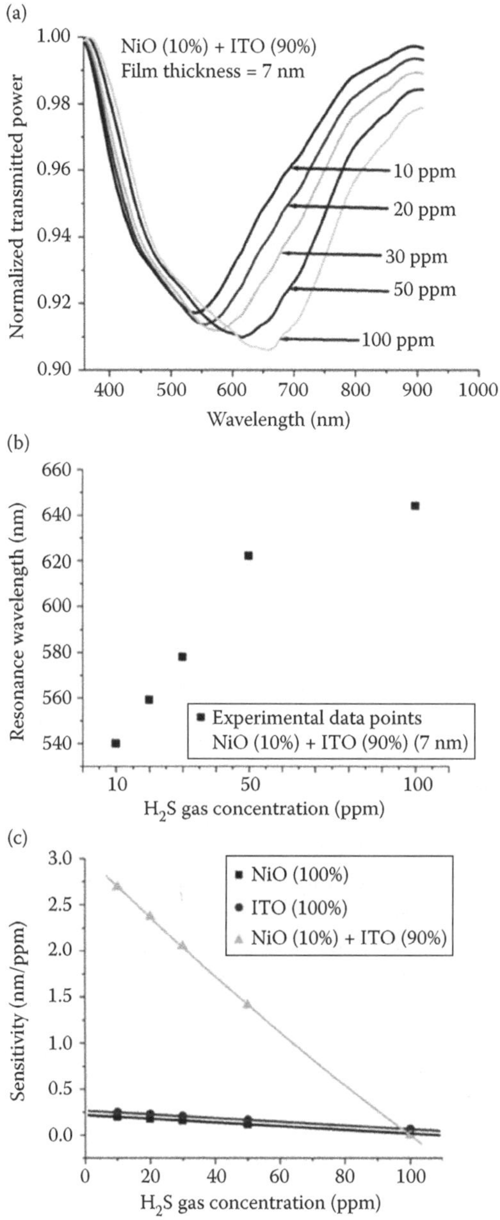

FIGURE 5.5
(a) SPR spectra, (b) calibration curve for H_2S gas sensor designed using NiO-doped ITO as the sensor layer, coated over silver film deposited on an unclad core of optical fiber, and (c) comparison of sensitivities of probe designed using NiO-doped ITO, only NiO, and only ITO as sensing layers. (Reprinted from *Sens. Actuators B* 195, Mishra, S. K. et al., Surface plasmon resonance based fiber optic hydrogen sulphide gas sensor utilizing nickel oxide-doped ITO thin film, 215–222. Copyright 2014, with permission from Elsevier.)

range of 10–20 ppm. As H_2S gas concentration increases, the amount of shift in resonance wavelength decreases. This was attributed to the limited availability of the active sites per gas molecule. This decreases as the concentration of the gas increases. Thus, the sensor holds a definite regime of operation.

For confirming that the chosen doped SMO material (NiO in ITO) is the best option for H_2S gas sensing, the experiments were repeated using two different probes. One, having coating of NiO and the other having ITO coating over silver-coated unclad fiber core. The thickness of both NiO and ITO layers was kept as 7 nm. The interaction of H_2S gas with probes having independent NiO and ITO layers converts them into nickel sulfide and ITO sulfide, respectively, as

$$NiO + H_2S \Leftrightarrow NiS + H_2O \tag{5.4}$$

$$ITO + H_2S \Leftrightarrow (ITO)S + H_2O \tag{5.5}$$

The formation of nickel sulfide and ITO sulfide over the sensing probe changes the refractive index of these sensing layers. Similar to the sensor probe having NiO-doped ITO, an increment in the H_2S gas concentration increases the refractive index of the sensing layer of both the probes. The behavior of calibration curves plotted for both the cases appears to be similar to the one shown in Figure 5.5b except with a smaller shift in resonance wavelength. The results show that an SMO doped with another SMO exhibits a higher number of defects and porousness than the individual SMOs, which supports a higher interaction with the gas molecules. A higher interaction results in a significant change in the refractive index of the sensing layer, making the sensor performance better. For all the three probes—NiO-doped ITO, only NiO, and only ITO—sensitivity values were calculated by taking the derivatives of the equations fitted to the calibration curves with respect to H_2S gas concentration. A maximum sensitivity of around 2.4 nm/ppm at 10 ppm concentration of H_2S gas was reported for the probe having NiO-doped ITO as the sensing element. A comparative plot showing sensitivity values evaluated for three probes is adapted from the work as shown in Figure 5.5c. The work demonstrated the realization of an SPR–SMO-based sensor with mechanism of sensing based on the chemical interaction of target gas on the doped SMO.

The process of doping in SMOs using noble metals to make sensor highly selective is a usual trend in SMO-based gas-sensing applications. As an example, a nanocomposite of palladium (Pd)-doped zinc oxide (ZnO) for hydrogen gas sensing was reported (Tabassum and Gupta 2015). The doping of noble metals in SMOs improves the electrical, electronic, and optical properties of SMOs which enhance their sensing behavior. The doping of palladium in SMO maximizes the sensitivity and makes the probe selective for hydrogen gas due to the ability of Pd ions to interact with the interstitial defects/

vacancies of SMO and occupies the space. Thus, the composite possesses extra native defects which enhance the interaction. In the work reported, a layer of nanocomposite, $ZnO_{(1-x)}Pd_x$, was coated over the silver-coated substrate (unclad core of the fiber) as the sensing element. The optimized value of factor "x" was obtained as 0.4. It corresponds to the maximum defects available for interaction with the gas. The adsorption of H_2 gas on nanocomposite, at room temperature, causes the hydration of Pd resulting in the crystallographic phase change of Pd from alpha (α) before adsorption to beta (β) after adsorption as shown in the following reaction:

$$Pd(\alpha) + \left(\frac{k}{2}\right) H_2 \Leftrightarrow PdH_k(\beta) \tag{5.6}$$

where k is the atomic ratio (H/Pd). The mechanism of sensing is depicted in Figure 5.6a, showing the phase transition of Pd and formation of palladium hydride. The formation of palladium hydride changes the effective refractive index of the sensing layer, which was confirmed from the shift of SPR spectrum on changing the concentration of H_2 gas. Since the lower explosion limit of H_2 gas at room temperature in atmosphere is 4% by volume, the work showed the sensor response for H_2 gas concentration from 0% to 4%.

The sensor probe fabricated is consisted of a thin film of silver having 40 nm thickness coated over 1 cm unclad portion of a long fiber piece followed by an overlayer of nanocomposite of thickness 94 nm coated using the dip coating method. A redshift in resonance wavelength and increase in depth of SPR curve were reported as shown in Figure 5.6b confirming the increment in real and imaginary parts of the effective refractive index of the sensing layer due to its interaction with 0% and 4% H_2. Palladium seems to induce substitutional defects on the ZnO surface which is used as the main property of hydrogen gas sensing in this work. The designed sensor probe was reported to have a good selectivity. Figure 5.6c shows the results of the selectivity test conducted on the probe showing the remarkable specificity to H_2 gas.

5.4.2 LMR-Based SMO Gas Sensors

Gas sensor designed with SMO utilizing LMR is the recent development in the field of environmental monitoring. The property of LMR exhibited by SMOs along with their gas-sensing property simplifies the sensor design and fabrication. In LMR–SMO-based gas sensor, the interaction of target gas with SMO changes its refractive index, which is transduced by the optical phenomenon of LMR. Similar to SPR–SMO-based sensors, the changes in the resonance wavelength and depth of LMR spectrum occur in LMR–SMO-based sensors due to the change in real and imaginary parts of the refractive index of SMO because of its interaction with gas molecules. SMO, to act as the sensing element, is chosen according to its interaction possibility with the target gas. Most of the SMOs support lossy modes when their film

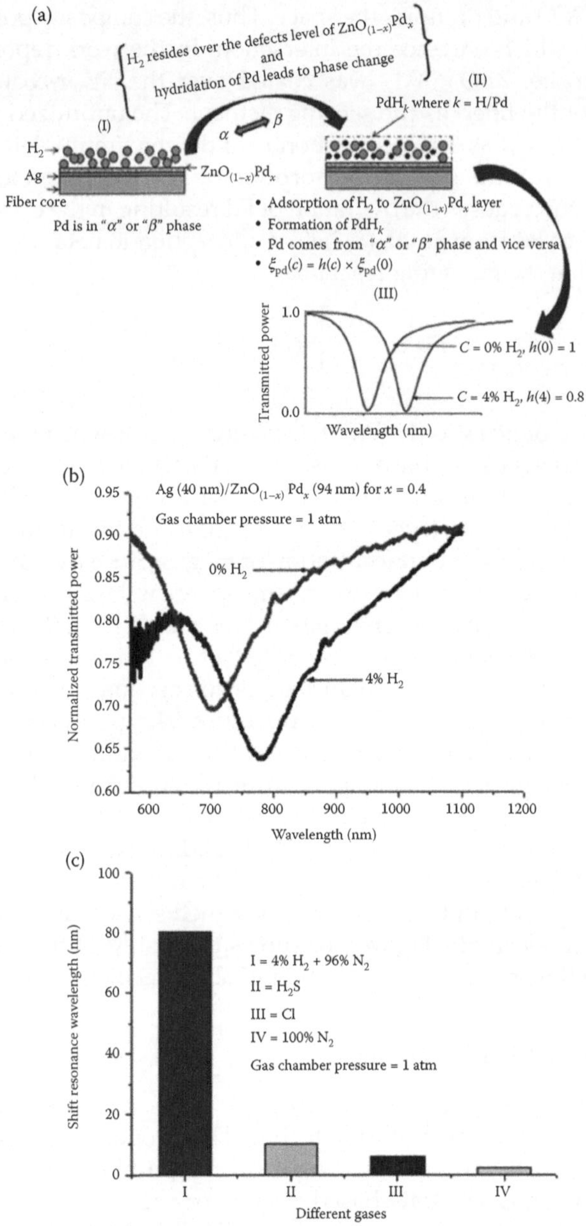

FIGURE 5.6
(a) Schematic of mechanism of H_2 sensing by the probe having palladium-doped ZnO as sensing layer, (b) SPR spectra recorded for the sensor probe characterized for 0% H_2 and 4% H_2, and (c) the selectivity test conducted on the probe showing the specificity to H_2 gas. (Reprinted from Tabassum, R. and B. D. Gupta. 2015. Fiber optic hydrogen gas sensor utilizing surface plasmon resonance and native defects of zinc oxide by palladium. *J. Opt.* 18: 015004. With permission from IOP Science.)

coated over an optical substrate is illuminated. The conditions for satisfying LMR, discussed in Chapter 2, must be followed by the design. A number of LMR–SMO-based bulk-layered and nanostructured gas sensors have been reported in the literature for environmental monitoring.

The comparison of gas sensors based on coatings of bulk layer, nanostructured, and bulk layer-nanostructured combination of SMO was reported (Usha et al. 2015a). The comparison was carried out on H_2S gas and the SMO chosen was zinc oxide (ZnO). Since the optical properties of a nanostructure are different from the bulk layer of materials like ZnO, its gas-sensing property is also expected to be different. The condition and method by which the SMO nanostructure is grown regulate its electrical, optical, and structural properties such as size and nature of defects. The role of ZnO bulk layer in exhibiting a high selectivity and response for a target gas is limited. Hence, the ZnO nanostructure that exhibits a high-aspect ratio (surface area to volume ratio) to provide better sensitivity, selectivity, stability, and compatibility with other devices and biological materials was used. The reported results also uphold the possibility of maximizing the sensor response by combining the bulk structure and nanostructure of ZnO. To verify and understand, three types of probes were fabricated, two based on LMR–SMO and the third based on SPR–SMO for a differential study as shown in Figure 5.7a through c. Figure 5.7a and b shows the LMR–SMO probe fabricated using both ZnO bulk + nanoparticle layer and only nanoparticle layer, respectively. SPR–SMO-based sensor prepared using silver film and ZnO bulk layer is shown in Figure 5.7c. The principle of sensing mechanism is based on the chemisorption and chemical reactions happening between the adsorbed oxygen species and defects with the dissociated gas molecules as explained in Section 5.4.

In the beginning, H_2S gas reacts with the adsorbed oxygen species (O^{2-} active sites) and releases H_2O and SO_2 according to the following reaction:

$$3O^{2-} + H_2S \Leftrightarrow SO_2 + H_2O \tag{5.7}$$

As the concentration of H_2S increases, the diffusion of H_2S gas into the structure increases and starts interacting with the interfacial/adsorbed oxygen defects (O^{2-}) and cations of zinc (Zn^{2+}) generated by oxygen species adsorption to generate SO_2 and ZnS, respectively, as shown in the following reactions:

$$ZnO \Leftrightarrow Zn^{2+} + O^{2-} \tag{5.8}$$

$$Zn^{2+} + O^{2-} + H_2S \Leftrightarrow ZnS + H_2O \tag{5.9}$$

These intermediates and by-products cover the surface of the sensing probe area, which changes the effective refractive index of the sensing layer. The shift in the LMR spectrum observed depicts this change in the

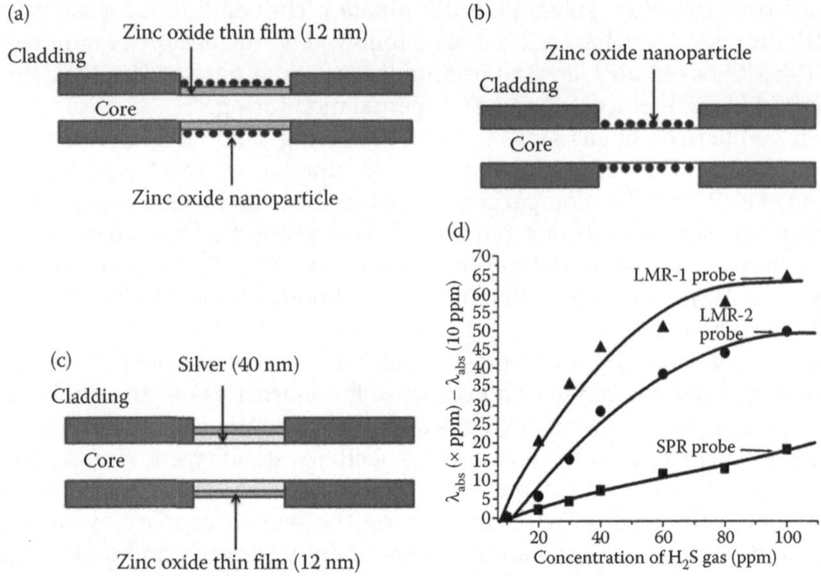

FIGURE 5.7
Schematic of LMR–SMO probe designed using (a) ZnO nanoparticle and bulk layer, (b) ZnO nanoparticle only, (c) SPR–SMO probe using silver and ZnO bulk layer for H_2S gas, and (d) comparison of variation in difference in peak absorbance or resonance wavelength evaluated from LMR or SPR response spectrum for each concentration of H_2S gas from its 10 ppm concentration. (Reprinted from *Sens. Actuators B* 218, Usha, S. P. et al., Fiber optic hydrogen sulfide gas sensors utilizing ZnO thin film/ZnO nanoparticles: A comparison of surface plasmon resonance and lossy mode resonance, 196–204. Copyright 2015, with permission from Elsevier.)

effective refractive index. Among the probes fabricated, LMR–SMO probe realized using combination of ZnO bulk layer of 12 nm thickness and nanoparticle layer, both deposited over 1-cm-long unclad core of the optical fiber, showed the best results. This is because of the higher interaction and diffusion possibility of the gas molecules in the case of combination of bulk layer and nanostructure of ZnO. The sensor probe was experimented for an H_2S gas concentration ranging from 10 to 100 ppm. The LMR and SPR spectra showed a redshift in peak absorbance wavelength (for LMR nanostructured probe) and resonance wavelength (for SPR probe) with the increase in H_2S gas concentration around the vicinity of the probe. The interaction of H_2S gas with ZnO results in the formation of zinc sulfide (ZnS) layer and the thickness of ZnS layer increases with the increase in the gas concentration. The increased thickness results in an increment in the effective refractive index of the sensing layer, which causes a shift in peak absorbance wavelength/resonance wavelength of the LMR/SPR spectrum. The refractive indices of ZnO and ZnS at 587.6 nm wavelength are 2.0034 and 2.3677, respectively.

Figure 5.7d shows the comparison of change in peak absorbance or resonance wavelength evaluated from LMR or SPR response spectrum for different concentrations of H_2S gas in reference to 10 ppm concentration of the gas. Commenting on the response of all the three sensor probes, the rate of change in peak absorbance/resonance wavelength decreases nonlinearly with the increase in H_2S gas concentration and becomes 0 for high concentration of H_2S gas. This is because of the decrement in the number of active sites available on the sensing surface and subsurface per gas molecule with the increase in the gas concentration. The probe with LMR–SMO design showed better results than SPR–SMO-designed probe. Moreover, a combination of bulk and nanostructure of ZnO seemed to provide a higher interaction feasibility for the target gas with the SMO structure. The probe so designed brought out a maximum shift in peak absorbance wavelength of 64 nm for H_2S gas concentration change from 10 to 100 ppm.

The gas–SMO interaction can be enhanced by providing a higher-surface area to volume ratio of the nanostructure. Using this concept, the work was later extended by replacing ZnO nanoparticles layer by a ZnO nanorod layer (Usha et al. 2015b). The sensor was fabricated by coating a 1 cm region of unclad core of a long fiber with a bulk layer of ZnO of 12 nm thickness. On this bulk layer, a thin layer of ZnO nanoparticles was dip coated using a hydrothermal route. ZnO nanorods were then grown from these nanoparticles seed layer. The sensor probe having ZnO nanorods grown for a period of 16 h seemed to support maximum interaction with H_2S gas. It was found that the nanorods grown for >16 h produced agglomerated rods limiting the area for gas interaction. The schematic of the sensor probe used for the experimentation is shown in Figure 5.8a.

The probe characterization for H_2S gas concentration from 10 to 100 ppm showed a total shift in peak absorbance wavelength of 83 nm. The peak absorbance wavelength extracted from the LMR spectra recorded for different concentrations of H_2S gas is plotted as a function of H_2S gas concentration in Figure 5.8b which is taken as the calibration curve. Increase in the aspect ratio in the case of nanorods compared to nanoparticles of ZnO increases the number of active sites that the gas molecules can easily outreach. This enhances the performance of the sensor in terms of total shift in peak absorbance wavelength by 1.2 times. The sensitivity determined using the derivative of the curve fitted to the calibration curve with respect to the H_2S gas concentration is also shown in the figure. The sensor probe showed a maximum sensitivity of 4.14 nm/ppm at 10 ppm H_2S gas concentration. The sensor was found to be suitable for H_2S gas concentration up to 60 ppm.

Employing the same idea, fiber-optic LMR–ITO-based H_2 gas sensor was reported (Mishra et al. 2016). The best performance of the sensor probe was obtained for coating of combination of ITO bulk layer and an overlayer of nanoparticles over the 1 cm unclad length of the core of the optical fiber as the sensing region. H_2 gas interacts with ITO forming ITO hydride causing a change in the refractive index of the sensing layer. The designed probe

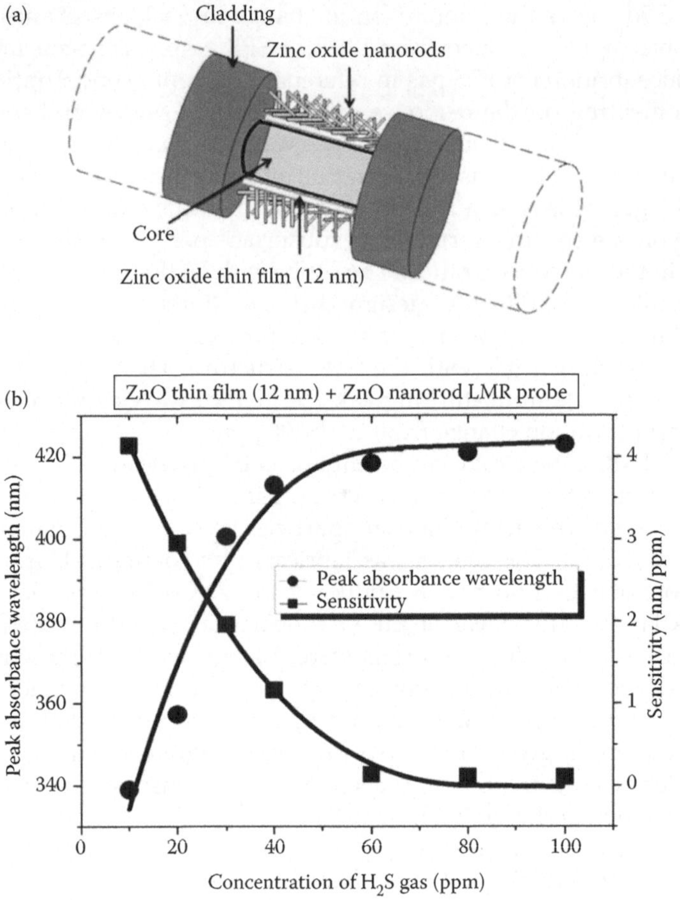

FIGURE 5.8
(a) Pictorial diagram of the LMR–SMO probe designed using ZnO nanorods and bulk layer and (b) variation of peak absorbance wavelength determined from LMR spectra and the sensitivity as a function of H_2S concentration. (Reprinted from Usha, S. P. et al. 2015. Zinc oxide thin film/nanorods based lossy mode resonance hydrogen sulphide gas sensor. *Mater. Res. Exp.* 2: 095003. With permission from IOP Science.)

operates for H_2 gas concentration in the range of 10–100 ppm, but the best response was obtained for a lower concentration range from 10 to 30 ppm. Figure 5.9a shows the variation of peak absorbance wavelength evaluated from the LMR spectra corresponding to different concentrations of H_2 gas. The sensor is highly sensitive to H_2 gas due to the chemical interactions between ITO and H_2 gas in addition to the interaction with the adsorbed oxygen species that may happen with any gas. The results of selectivity of the probe for various other gases for the change in concentration from 10 to 100 ppm are shown in Figure 5.9b.

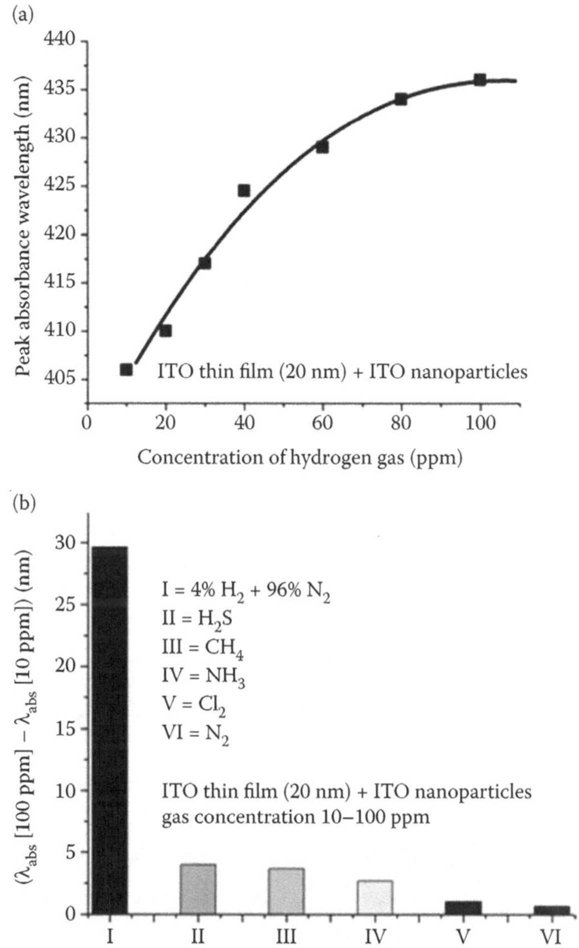

FIGURE 5.9
(a) Calibration curve showing the peak absorbance wavelength evaluated from the LMR spectrum recorded for H_2 gas concentration of 10–100 ppm and (b) selectivity graph plotted showing total shift in peak absorbance wavelength evaluated for various gases for the sensor probe designed with a combination of ITO thin film and nanoparticles as the sensing layer. (Reprinted from Mishra, S. K. et al. 2016. A lossy mode resonance based fiber optic hydrogen gas sensor for room temperature using coatings of ITO thin film and nanoparticles. *Meas. Sci. Technol. B* 27: 045103. With permission from IOP Science.)

5.5 Biosensing with SMOs

The nanostructures of SMOs have also received much attention for biomedical applications. In comparison to bulk structure, nanostructure provides higher effectual surface so that any biomolecule can be easily immobilized with better emplacement giving its maximal biological activity. SMO in nanostructured forms also sustains perfect interfacing between biorecognition element and the transducer component. Biosensors utilizing SMOs possess the advantage of enhancement of sensor's performance through SMO's biocompatibility, adjustable morphology to give adaptability, surficial interactivity, catalytic behavior, and direct adsorption/immobilization properties to attach the maximum biomolecules (Solanki et al. 2011). These benefits make SMO-based biosensors more compatible for both biomedical and clinical applications. Figure 5.10 shows a brief classification of the common SMOs used for biosensing. The classification is based on the property of SMO that supports the biomedical sensing.

For biosensing, the selection of SMO is critical since it decides the sensor realization and practicality. The SMO should provide easy immobilization of biomolecule and the interface between them should be good enough to make the sensor stable. Usually, the interaction between SMO and biomolecule is either physical or chemical in nature. Physical interaction can be physisorption, electrostatic attraction, or van der Waals interaction depending on the SMO surface charge and adsorption mechanism. The

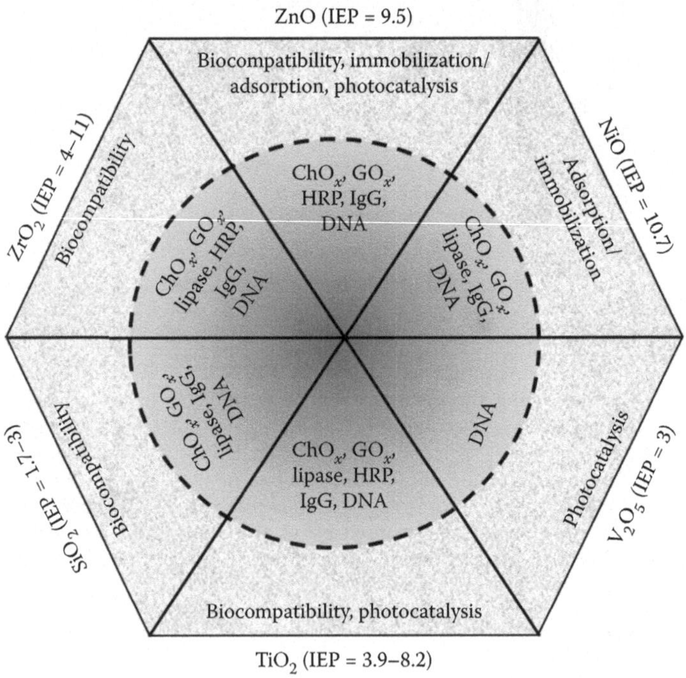

FIGURE 5.10
Common SMOs used for biosensing applications with their isoelectric points, primary properties supporting the biosensing, and the common biomolecules/species that are immobilized directly on SMO.

chemical interaction can be covalent binding, cross-linking, or any other which depends on functionalities available to bind with the biomolecule. The fabrication route of SMO nanostructure and combinational structure also plays an important role in controlling the sensing properties. Most of the SMOs used are, generally, of tin, zinc, iron, magnesium, nickel, and silicon. Their structures include nanoparticles, nanorods, core shell, and nanofibers. The incorporation of nanostructures of highly conducting materials such as graphene, silver, and carbon nanotube in SMOs has also been reported to enhance the electrical and optical properties of SMOs. Hence, such combinations are also preferred to enhance the electron transfer mechanism and the biosensing properties. Few examples of biosensing with SMO are discussed in this section. The examples chosen are for the detection of insulinoma and D-sorbitol.

A body disorder of low blood dextrose from its normal range is called hypoglycemia. Insulinoma is a dangerous cause to hypoglycemia, which is usually made conclusive by Whipple's triad. It is a 72 h test with a prefixed meal including the monitoring of dextrose (D-glucose) of the patient. Every 6 h the patient's blood needs to be tested and is completed when the patient starts showing the symptoms of hypoglycemia with the body sugar level <2.2 mM. A biosensor using nanostructured SMO for insulinoma monitoring application by measuring the blood dextrose has been reported in the literature (Usha et al. 2016). The sensor probe designed on the optical fiber implementing the technique of SPR was fabricated using silver, zinc oxide nanorods, and glucose oxidase (GO_x). Silver film of 40 nm thickness was coated over unclad core of the fiber to realize SPR. The zinc oxide nanorods (nanostructured SMO here) were grown hydrothermally from the seed layer of zinc oxide nanoparticles over silver film to play a dual function. Due to the high-refractive index layer over silver film, ZnO nanostructure enhances the sensor probe performance in terms of sensitivity. Further, it also acts as a matrix for the direct immobilization of the enzyme GO_x. The direct immobilization of enzyme was performed using high adsorption by exploiting the difference in the isoelectric points of the zinc oxide (9.5) and glucose oxidase (4.2). The SEM image of the finalized sensor probe showing the glucose oxidase immobilized over ZnO nanorods is shown in Figure 5.11a.

The sensor probe was characterized for the dextrose concentration ranging from 0 to 10 mM. The series of reactions involved in sensing of dextrose in the presence of enzyme GO_x are shown in Figure 5.11b. GO_x is an enzyme that comes in the family of oxidoreductase, which undergoes both reduction and oxidation process while interaction. D-glucose reacts with GO_x in the presence of oxygen and water to form D-gluconolactone and hydrogen peroxide (H_2O_2). These intermediate products possess higher refractive index than GO_x and dextrose causing an increment in the effective refractive index of the sensing layer. Initially, the flavin group (FAD) of the enzyme undergoes reduction in the presence of D-glucose to form

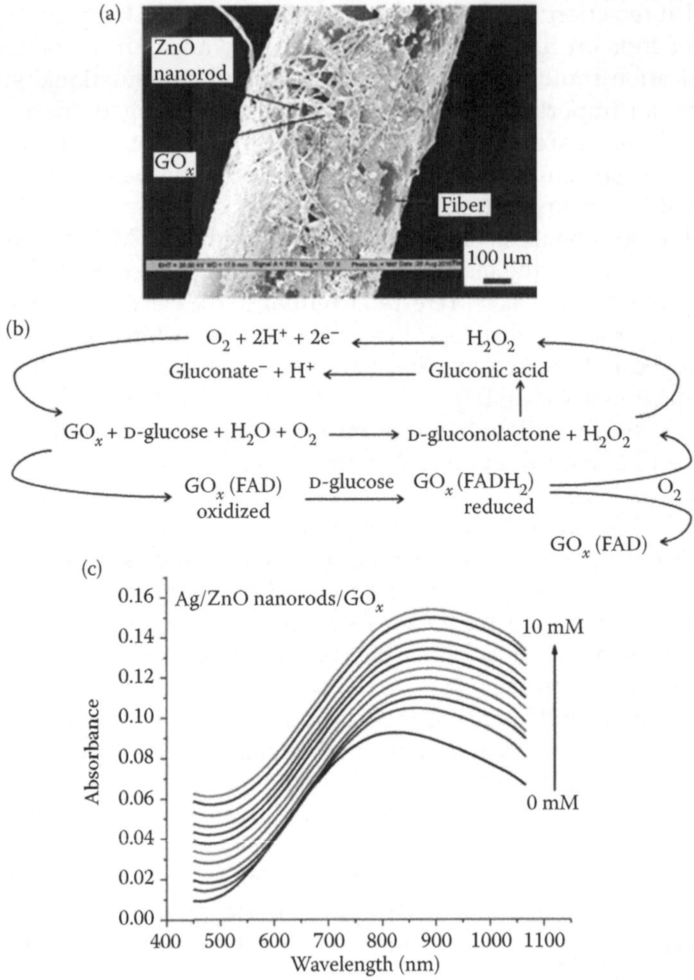

FIGURE 5.11
(a) SEM micrograph of the finalized sensor probe showing the glucose oxidase immobilized over ZnO nanorods shown over optical fiber, (b) series of reactions between D-glucose and glucose oxidase enzyme to change the effective refractive index of the sensing layer, and (c) the SPR spectra recorded with change in glucose concentration in the range of 0–10 mM. (Reprinted from *Biosens. Bioelectron.* 85, Usha S. P. et al., FO-SPR based dextrose sensor using Ag/ZnO nanorods/GO_x for insulinoma detection, 986–995. Copyright 2016, with permission from Elsevier.)

$FADH_2$. Reduction is followed by the reoxidation of $FADH_2$ to FAD. During the reoxidation process, the flavin group accepts oxygen molecule and generates H_2O_2, that dissociates to H^+ ions and electrons. At the same time, D-gluconolactone hydrolyses to form gluconic acid, which dissociates to gluconate$^-$ and H^+ ions. All the reactions appeared to be reversible generating back the oxidized form of GO_x to interact with the D-glucose in the presence of necessary oxygen to carry forward the reactions while the experiments were repeated.

The reaction between glucose oxidase and dextrose producing the intermediate and by-products changes the refractive index of the sensing layer. The SPR response spectra showed an increase in the value of absorbance and a redshift in the peak absorbance wavelength with the increase in the dextrose concentration as shown in Figure 5.11c. The response of the probe was reported to be quite fast (response time = 54 s). In this sensor, the biocompatibility and direct immobilization properties of ZnO have made the sensor probe fabrication simpler and more stable for biomedical application.

Another example of biosensor that uses SMO nanostructure is for the sensing of D-sorbitol. D-sorbitol is a clinically important analyte and is used as a preservative in food industry. Its excess intake may result in abdominal pain. A sensor developed for D-sorbitol sensing utilizes enzyme sorbitol dehydrogenase (SDH) embedded in nanoflowers of tantalum pentaoxide (Ta_2O_5) (Kant et al. 2017). Ta_2O_5 nanoflowers provide a high-index layer as well as entrapment of the enzyme. SPR–SMO design was implemented by coating a 40-nm-thick layer of silver over unclad core region of the optical fiber. Over the silver film, the enzyme entrapped Ta_2O_5 nanoflowers were deposited using dip coating technique. The sensor probe was characterized for a D-sorbitol concentration range of 0.1–1.1 μg/mL. In the prepared sample solutions of D-sorbitol, the coenzyme of SDH, nicotinamide adenine dinucleotide (NAD), was also added to release NAD^+ to catalyze the reaction. The reaction between D-sorbitol and SDH in the presence of catalyst NAD^+ is shown below.

$$\text{D-sorbitol} + \text{SDH} + \text{NAD}^+ \Leftrightarrow \text{D} - \text{fructose} + \text{NADH} + \text{H}^+ \qquad (5.10)$$

The formation of complex by-products by the interaction of SDH and D-sorbitol in the presence of NAD^+ covers the sensing probe region and changes the effective refractive index of the sensing layer. A blueshift in resonance wavelength was observed with the increase in the concentration of D-sorbitol. For the change in concentration from 0.1 to 1.1 μg/mL, a shift of around 55 nm in resonance wavelength was observed. A decrement in the resonance wavelength with the increase in D-sorbitol concentration was due to the decrease in the real part of the effective refractive index of the sensing layer. The work also showed the influence of morphology of Ta_2O_5 in entrapping the enzyme. Figure 5.12 shows the comparative study conducted by entrapping the enzyme in different ways such as using Ta_2O_5 thin film and nanoparticles in addition to Ta_2O_5 nanoflowers. In this sensor, the topographical features of nanoflowers seem to support a higher-enzyme loading and hence the complex formation by providing a greater aspect ratio in comparison to other morphologies.

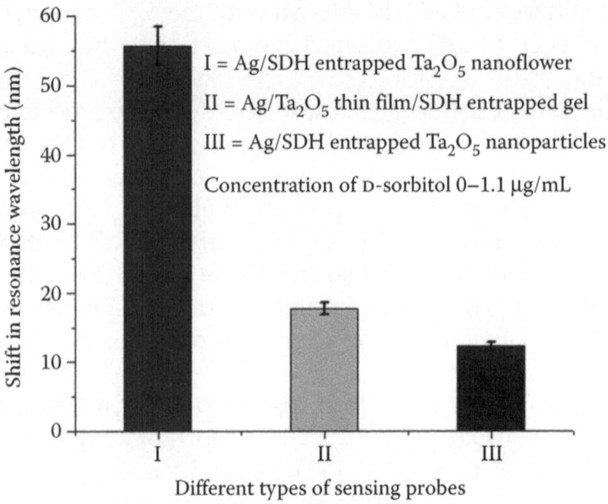

FIGURE 5.12
Results showing the comparative study conducted by entrapping the enzyme in different ways such as using Ta_2O_5 thin film and nanoparticles in addition to Ta_2O_5 nanoflowers. (Reprinted from *Sens. Actuators B* 242, Kant, R., A highly sensitive and distinctly selective D-sorbitol biosensor using SDH enzyme entrapped Ta_2O_5 nanoflowers assembly coupled with fiber optic SPR, 810–817. Copyright 2017, with permission from Elsevier.)

5.6 Summary

The chapter summarizes the role of SMOs in sensing/monitoring of environmental parameters and biomedical/clinical analytes. The properties of SMO to exhibit optical transparency and conductivity in visible and infrared regions, stability and mechanical flexibility, pressure tolerance, and compatibility with inorganic and organic materials make them a distinctive material in sensing applications. This chapter discusses various factors which change the conductometric property of SMOs. For gas-sensing applications, the change in the electrical resistance/conductivity of SMOs on exposure to oxidizing and reducing gases is used. The interaction of oxidizing and reducing gases by adsorption or chemical interaction with SMOs changes the surface chemistry of metal oxides having surface and native oxygen defects which change the effective refractive index of the metal oxide. Due to these reasons, SMOs have been used for gas sensing utilizing SPR/LMR technique. In addition to bulk layered SMO, the feasibilities and advantages of incorporating the nanostructured SMOs in sensing have also been outlined. Since SMO has been the major component of gas sensor probe, a detailed discussion on both SPR- and LMR-based SMO mechanisms has been included. In the last section, we have briefly discussed the use of nanostructured SMOs in realizing the biosensors. Some of the gas sensors and biosensors have been discussed that use bulk and nanostructures of SMOs and utilize SPR/LMR technique. These sensors can be used for environmental/gas monitoring and clinical applications.

References

Anta, J. A. 2012. Electron transport in nanostructured metal-oxide semiconductors. *Curr. Opin. Colloid Interface Sci.* 17: 124–131.

Arafat, M. M., B. Dinan, S. A. Akbar, and A. S. M. A. Haseeb. 2012. Gas sensors based on one dimensional nanostructured metal-oxides: A review. *Sensors* 12: 7207–7258.

Barsan, N., D. Koziej, and U. Weimar. 2007. Metal oxide-based gas sensor research: How to? *Sens. Actuators B* 121: 18–35.

Bielański, A. and M. Najbar. 1972. Adsorption species of oxygen on the surfaces of transition metal oxides. *J. Catal.* 25: 398–406.

Brinzari, V., G. Korotcenkov, and V. Golovanov. 2001. Factors influencing the gas sensing characteristics of tin dioxide films deposited by spray pyrolysis: Understanding and possibilities of control. *Thin Solid Films* 391: 167–175.

Dawar, A. L. and J. C. Joshi. 1984. Semiconducting transparent thin films: Their properties and applications. *J. Mater. Sci.* 19: 1–23.

Dutronc, P., C. Lucat, F. Menil, M. Loesch, and L. Combes. 1993. A new approach to selectivity in methane sensing. *Sens. Actuators B* 15: 24–31.

Fine, G. F., L. M. Cavanagh, A. Afonja, and R. Binions. 2010. Metal oxide semi-conductor gas sensors in environmental monitoring. *Sensors* 10: 5469–5502.

Henrich, V. E. and P. A. Cox. 1996. *The Surface Science of Metal Oxides*. Cambridge: Cambridge University Press.

Jia, N. Q., J. Xu, M. H. Sun, and Z. Y. Jiang. 2005. A mediatorless hydrogen peroxide biosensor based on horseradish peroxidase immobilized in tin oxide sol-gel film. *Anal. Lett.* 38: 1237–1248.

Kant, R., R. Tabassum, and B. D. Gupta. 2017. A highly sensitive and distinctly selective d-sorbitol biosensor using SDH enzyme entrapped Ta_2O_5 nanoflowers assembly coupled with fiber optic SPR. *Sens. Actuators B* 242: 810–817.

Kaushik, A., R. Khan, P. R. Solanki et al. 2008. Iron oxide nanoparticles–chitosan composite based glucose biosensor. *Biosens. Bioelectron.* 24: 676–683.

Korotcenkov, G. 2005. Gas response control through structural and chemical modification of metal oxide films: State of the art and approaches. *Sens. Actuators B* 107: 209–232.

Korotcenkov, G. 2007. Metal oxides for solid-state gas sensors: What determines our choice? *Mater. Sci. Eng. B* 139: 1–23.

Korotcenkov, G., S. V. Dmitriev, and V. I. Brynzari. 1999. Processes development for low cost and low power consuming SnO_2 thin film gas sensors (TFGS). *Sens. Actuators B* 54: 202–209.

Korotcenkov, G., V. Golovanov, V. Brinzari, A. Cornet, J. Morante, and M. Ivanov. 2005. Distinguishing feature of metal oxide films structural engineering for gas sensor applications. *J. Phys. Conf. Ser.* 15: 256–261.

Korotcenkov, G., V. Macsanov, V. Tolstoy, V. Brinzari, J. Schwank, and G. Faglia. 2003. Structural and gas response characterization of nano-size SnO_2 films deposited by SILD method. *Sens. Actuators B* 96: 602–609.

Liu, A., M. D. Wei, I. Honma, and H. Zhou. 2006. Biosensing properties of titanate nanotube films: Selective detection of dopamine in the presence of ascorbate and uric acid. *Adv. Funct. Mater.* 16: 371–376.

Liu, S., Z. Dai, H. Chen, and H. Ju. 2004. Immobilization of hemoglobin on zirconium dioxide nanoparticles for preparation of a novel hydrogen peroxide biosensor. *Biosens. Bioelectron.* 19: 963–969.

Mishra, S. K., S. Rani, and B. D. Gupta. 2014. Surface plasmon resonance based fiber optic hydrogen sulphide gas sensor utilizing nickel oxide doped ITO thin film. *Sens. Actuators B* 195: 215–222.

Mishra, S. K., S. P. Usha, and B. D. Gupta. 2016. A lossy mode resonance based fiber optic hydrogen gas sensor for room temperature using coatings of ITO Thin film and nanoparticles. *Meas. Sci. Technol.* 27: 045103.

Pathak, A., S. K. Mishra, and B. D. Gupta. 2015. Fiber optic ammonia sensor using Ag/SnO_2 thin films: Optimization of thickness of SnO_2 film using electric field distribution and reaction factor. *App. Opt.* 54: 8712–8721.

Peter, Y. U. and M. Cardona. 2010. *Fundamentals of Semiconductors: Physics and Material Properties.* Heidelberg: Springer.

Salimi, A., E. Sharifi, A. Noorbakhsh, and S. Soltanian. 2007. Immobilization of glucose oxidase on electrodeposited nickel oxide nanoparticles: Direct electron transfer and electrocatalytic activity. *Biosens. Bioelectron.* 22: 3146–3153.

Singh, S. and B. D. Gupta. 2010. Simulation of surface plasmon resonance based fiber optic sensor for gas sensing in visible range using films of nanocomposites. *Measure. Sci. Technol.* 21: 115202.

Solanki, P. R., A. Kaushik, V. V. Agrawal, and B. D. Malhotra. 2011. Nanostructured metal oxide-based biosensors. *NPG Asia Mater.* 3: 17–24.

Tabassum, R. and B. D. Gupta. 2015. Fiber optic hydrogen gas sensor utilizing surface plasmon resonance and native defects of zinc oxide by palladium. *J. Opt.* 18: 015004.

Usha, S. P., S. K. Mishra, and B. D. Gupta. 2015a. Fiber optic hydrogen sulfide gas sensors utilizing ZnO thin film/ZnO nanoparticles: A comparison of surface plasmon resonance and lossy mode resonance. *Sens. Actuators B* 218: 196–204.

Usha, S. P., S. K. Mishra, and B. D. Gupta. 2015b. Zinc oxide thin film/nanorods based lossy mode resonance hydrogen sulphide gas sensor. *Mater. Res. Exp.* 2: 095003.

Usha, S. P., A. M. Shrivastav, and B. D. Gupta. 2016. FO-SPR based dextrose sensor using Ag/ZnO Nanorods/GO_x for insulinoma detection. *Biosens. Bioelectron.* 85: 986–995.

Xu, J. J., X. L. Luo, Y. Du, and H. Y. Chen. 2004. Application of MnO_2 nanoparticles as an eliminator of ascorbate interference to amperometric glucose biosensors. *Electrochem. Comm.* 6: 1169–1173.

Yu, X., T. J. Marks, and A. Facchetti. 2016. Metal oxides for optoelectronic applications. *Nat. Mater.* 15: 383–396.

6

Molecular-Imprinting-Based Sensors

The sensing technology, as mentioned in previous chapters, basically depends on two essential elements: recognition unit and transducing unit. The recognition unit (or sensing layer) changes its properties with the interaction of the target analyte/template. The change is converted into readable signal using the transducer unit and can be analyzed by means of the detector. There are a number of naturally occurring receptors which are extensively used in the sensor fabrication such as antibodies, enzymes, amino acids, DNAs, etc. These receptors are mostly selective for the detection of the template molecules but they suffer from numerous disadvantages such as high cost, complex preparation method, instability in the harsh atmosphere, and not having freedom to choose the template (Subrahmanyam et al. 2002). Thus, to overcome these disadvantages, the work on the synthesis of artificial receptors was pioneered and for this, Cram, Lehn, and Pederson won the Nobel Prize in 1987 (Lehn 1995). The greatest advantage of the synthetic receptors is that these have the freedom of molecular design, that is, the variety of templates can be used for the preparation of artificial receptors. The designing of artificial receptors with high specificity, easy synthesis, and high stability is always a great challenge for the researchers (Amabilino and Stoddart 1995). In the era of fabrication of artificial receptors, molecular imprinting came up in which the highly selective memory encampments of the template are fabricated in a polymeric medium. These encampments work as the recognition sites for the template detection; the polymers with template-imprinted sites are called molecularly imprinted polymers (MIPs) and the technology is termed as molecular-imprinted technology (MIT). The process of MIP synthesis includes the trapping of template in a polymerized medium of a functional monomer. The template molecules are then removed from the polymerized medium with a suitable remover. The removal of template molecules from the polymer results in the creation of 3D empty space with opposite shape and size to the template in the polymer. This empty space is known as the recognition site for the template molecule within the polymer (Chen et al. 2016). As the recognition sites have complementary imprints of the template, these are highly selective to interact with the specific template molecule, which is the most important feature of MIT. Besides this, MIPs have numerous other advantages of easy synthesis procedure, high stability even at harsh environments, remarkable robustness, and low cost. Considering these universal features, MIPs attract the researchers and the scientific community to develop the template-specific polymers and apply these in number of applications such as solid-phase extraction, chiral discrimination, chemical/biosensing, preparation of artificial antibodies, catalysis, and degradation (Chen et al. 2011).

This chapter provides a brief discussion on the role of molecular imprinting in optical sensing and its advantages. The chapter starts with a brief introduction of the fundamentals of MIPs with the role of each element and the synthesis procedures for the fabrication of these synthetic polymers. In Section 6.2, the types of molecular imprinting based on the interaction with analyte have been discussed with their advantages and disadvantages. Section 6.3 is devoted to the use of molecular imprinting as the sensing element by means of various optical methods and its advantages over the other conventional detection schemes. Section 6.4 is about the numerous studies reported in the literature for the application of MIPs in the areas of food safety, environmental control, and biomedical applications. The final section of the chapter summarizes the role of molecular imprinting in optical sensing technology with their pros, cons, and the future possibilities.

6.1 Basics of Molecular Imprinting

The procedure for the synthesis of molecular-imprinted polymers requires the freezing of template molecule in a polymeric atmosphere which is achieved via the polymerization of the functional monomer. In the process, initially, a prepolymerized solution is prepared with the cross-linking of template and monomer with the help of a suitable cross-linking agent, solvent, and the initiator. The prepolymerized template-monomer solution is further polymerized under certain polymerization atmospheric condition (suitable temperature, pressure, etc.). This results in the creation of the polymeric medium having template molecules within its vicinity, which is termed as the nonimprinted polymer (NIP). The next step of molecular imprinting involves the removal of template molecules from the vicinity of the NIP. In this step, polymeric medium is washed with a required remover such that it unbinds/removes the interaction between functional monomer and template molecules. After removing, the template molecules leaves their imprints in the polymer as a 3D vacant space with complementary shape and size of the template. As discussed earlier, due to the complementary imprints to the templates, these are highly selective for further recognition of template molecules. Figure 6.1 shows the pictorial representation of the step-by-step synthesis of the molecular-imprinted polymer and the reason behind its highly selective nature.

The performance of MIP is evaluated by means of various parameters such as stability, optimized porosity, rigidity, specificity, etc. These are strongly dependent on the elements and synthesis procedures used for MIP

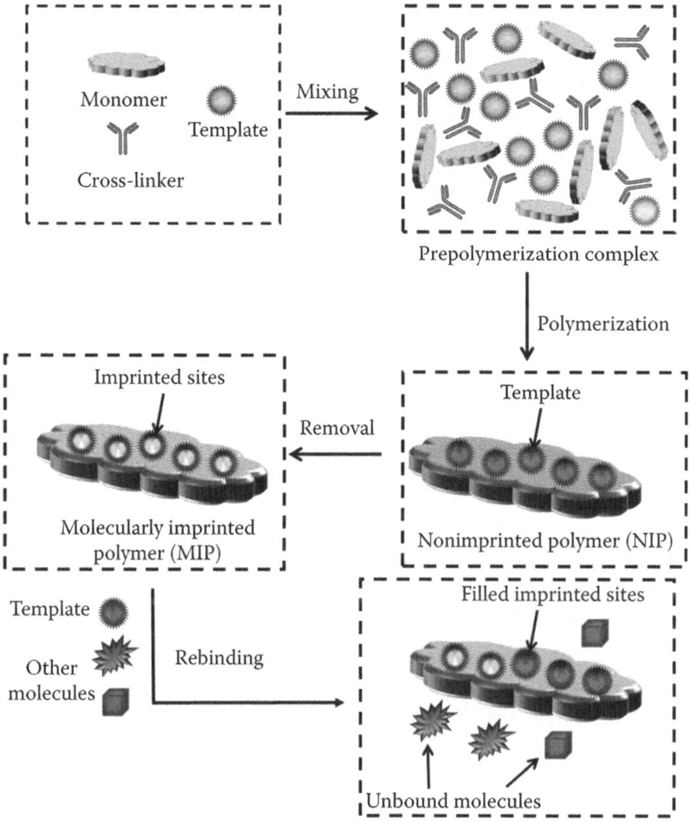

FIGURE 6.1
Schematic of the step-by-step preparation of a generic molecular-imprinted polymer with highly selective recognition of the template.

fabrication. The brief discussion on these has been provided in the following sections.

6.1.1 Molecular-Imprinting Elements

There are five basic elements which are usually used for the fabrication of MIPs: template, monomer, cross-linking agent, reaction initiator, and solvent. The properties of the MIP synthesized can easily be tailored by varying the types and quantities of monomer, cross-linker, reaction initiator, and solvent along with the polymerization temperature and period. We discuss each element below.

Template: Template is the target molecule whose imprints are to be prepared as the recognition sites in MIP. The size of the template may vary from atom, ion, molecule, molecular complex, and ionic compound to macromolecule. This is the main advantage of molecular imprinting as it provides the freedom of selecting the variety of templates. There are three main essential conditions for an ideal template: (i) template should have the functional groups which can help the formation of complexes with the monomers, (ii) it should not contain functional group which restricts the complex formation with monomer, and (iii) template should exhibit a worth chemical stability during polymerization (Chen et al. 2011). In the case of metal ion as the template, the prepared imprinted polymer is, generally, termed as ion-imprinted polymer (IIP). In the synthesis procedure for IIP, initially, the metal ion is complexed with a certain ligand molecule which is then polymerized with the functional monomer. After polymerization process, the elution of metal ion is performed for the creation of ion-imprinted sites within the polymer. Till now, a variety of templates have been used for the synthesis of MIPs, which include organic molecules (amino acids, sugars, cells, pharmaceuticals, etc.), various ions, biomolecules, cells, etc. (Chen et al. 2011). However, the imprinting of large molecules such as proteins is possible but their efficient imprinting is still a challenge due to their large size (Zhang et al. 2011). Table 6.1 summarizes some of the template molecules.

Functional monomers: The use of functional monomer is to provide the functional group which makes the prepolymerization complex by interacting with the template. Thus, it is essential to use suitable monomer which can provide a strong interaction with the template. The interaction between template and monomer may be covalent or noncovalent. Based on these interactions, MIPs are divided into two categories: covalent imprinting and noncovalent imprinting. Former category corresponds to the template–monomer interaction which is covalent, while noncovalent imprinting corresponds to the noncovalent template–monomer interaction. The properties of these types of imprinting will be discussed in Section 6.2. Thus, the selection of the monomers is quite necessary as it strongly affects the properties of the polymers. Methacrylic acid (MAA) is one of the mostly used

TABLE 6.1

Typical Templates Used for MIP Synthesis

Type	Examples
Medicines	Erythromycin, tetracycline, propranolol
Amino acids and peptides	Alanine, tyrosine, helical peptide, tripeptide
Sugars	Glucose, dextrose, fructose
Pesticides	Profenofos, atrazine, benzimidazole fungicides
Explosives	2,4,6-Trinitrotoluene (TNT)
Endocrine-disrupting chemicals (EDC)	Estradiol, bisphenol A
Viruses	Dengue virus, bovine leukemia virus, tobacco mosaic virus
Proteins	Bovine serum albumin (BSA)

monomers for MIP synthesis and known as the universal functional monomer. This is because of its ability for the creation of hydrogen bond with various functional groups. In addition, while polymerization the dimerization of MAA occurs which also significantly improves the efficiency of MIPs (Zhang et al. 2010). Further, higher-molar concentrations of MAA causes the increase in the pore structure of the MIP which results in the enhanced binding capacity. Table 6.2 represents common functional monomers used in MIP synthesis.

Apart from monomer type, the quantity of the monomer also strongly affects the performance of MIP such as number, affinity, and stability of imprinted sites. The monomer quantity required for an effective imprinting depends upon its ratio to the quantity of template used (Chen et al. 2016). The low value of monomer to template ratio (M:T) causes the formation of less number of the monomer–template complex due to the inadequate amount of monomer. Hence, less number of imprinted sites is created and results in the minimal binding ability. On the other hand, for higher M:T value, few monomers are retained within the prepolymerized solution after the formation of the complete monomer–template complex. Thus, retained monomers will cause the unspecific binding with other interfering analytes and a decrease in the selectivity of MIP. To increase the performance of the MIP, there is a possibility to synthesize the monomer, according to its application. The synthesis procedure for this is tuned according to the corresponding template.

Cross-linker: The cross-linker is used to provide bonding between the functional groups of monomer and template. The cross-linking of template and monomer provides the effective rigid/porous structure to the imprinted

TABLE 6.2

Typical Monomers Used for MIP Synthesis with Different Template: Monomer Interactions

	Monomer	Chemical Formula
Covalent	4-Vinyl benzene boric acid	$H_2C{=}CHC_6H_4B(OH)_2$
	4-Vinyl benzaldehyde	C_9H_8O
	4-Vinyl aniline	$H_2C{=}CHC_6H_4NH_2$
Noncovalent	Acrylic acid (AA)	$CH_2{=}CHCOOH$
	Methacrylic acid (MAA)	$H_2C{=}C(CH_3)COOH$
	Trifluoro methacrylic acid (TFMAA)	$H_2C{=}C(CF_3)CO_2H$
	Methyl methacrylate (MMA)	$CH_2{=}C(CH_3)COOCH_3$
	Itaconic acid	$HO_2CCH_2C({=}CH_2)CO_2H$
	Styrene	$C_6H_5CH{=}CH_2$
	4-Vinylpyridine	C_7H_7N
	Acrylamide (AM)	$CH_2{=}CHCONH_2$
	2-Hydroxyethyl methacrylate (HEMA)	$CH_2{=}C(CH_3)COOCH_2CH_2OH$
Semicovalent	3-Isocyanatopropyl tetraethoxysilane (IPTS)	$(OC_2H_5)_3SiNCO(CH_2)_3$

polymers even after removing the template from NIP. The type and quantity of the cross-linking agent also affect the selectivity and binding capability of the MIP. The small amount of cross-linker results in poor rigidness or higher flexibility which affects the stability of binding sites structure causing less selectivity. It is found that the selective nature of MIP is observed for the least cross-linking degree of 10% (Wulff et al. 1982), while a higher amount of cross-linker causes the synthesized MIP with restrictions for binding toward template recognition, as it decreases the number of recognition sites per unit mass of MIP. Hence, to achieve the best performance of MIP, one needs the suitable type and amount of the cross-linker. The most commonly used cross-linkers are ethylene glycol dimethacrylate (EGDMA), divinylbenzene (DVB), and trimethylolpropane trimethacrylate (TRIM). EGDMA is the most suitable cross-linker when the monomers from an acrylic group such as MAA are used. The synthesized polymers are not only highly selective but also exhibit good stability, porosity, and wettability. In the case of TEGDMA (tri-EGDMA) as cross-linker, its hydrophilic nature increases the number of the binding sites for the template recognition. DVB is used as the cross-linker when MIP is synthesized using polystyrene. Few other examples of the cross-linkers used are shown in Table 6.3. Like monomers, the applications of cross-linkers are restricted due to their limited molecular structure. Hence, for tuning the flexibility/rigidity of MIPs, synthetic cross-linkers can also be prepared.

TABLE 6.3

Typical Cross-Linkers Used for the MIP Fabrication Methods with Different Template: Monomer Interactions

	Cross-Linker	Chemical Formula
Covalent	Bis-(1-(tert-butylperoxyl)-1-methylethyl)-benzene (BIPB)	$C_6H_4[C(CH_3)_2OOC(CH_3)_3]_2$
	Dicumyl peroxide (DCP)	$[C_6H_5C(CH_3)_2]_2O_2$
	Triallyl isocyanurate (TAIC)	$C_{12}H_{15}N_3O_3$
Noncovalent	Ethylene glycol dimethacrylate (EGDMA)	$CH_2{=}C(CH_3)COOCH_2CH_2OCOC(CH_3){=}CH_2$
	Divinylbenzene (DVB)	$C_6H_4(CH{=}CH_2)_2$
	Glycidyl methacrylate (GMA)	$C_7O_3H_{10}$
	Trimethylolpropane trimethacrylate (TRIM or TMPTM)	$C_{18}O_6H_{26}$
	1,3-Diisopropyl benzene	$C_6H_4[CH(CH_3)_2]_2$
Used in sol-gel	Tetramethoxysilane (TMOS)	$Si[OCH_3]_4$
	Tetraethoxysilane (TMOS)	$Si[OC_2H_5]_4$
	Phenyltriethoxysilane (PTEOS)	$C_6H_5Si[OC_2H_5]_3$
	Diphenyldiethoxysilane (DPDES)	$[C_6H_5]_2Si[OC_2H_5]_2$

Polymerization initiators: Molecular imprinting is broadly performed by free radical polymerization, electropolymerization, and photopolymerization methods. The rate of free radical polymerization is strongly affected by the initiator properties (Lanza et al. 2001). 2-Azobisisobutyronitrile (AIBN) is the mostly used initiator for polymerization. This is because of thermal decomposition of AIBN which results in the formation of two isobutyronitrile radicals, which initiate the polymerization process. For free radical polymerization, mostly azo-group compounds are used as the initiators. Figure 6.2 shows the few initiators used for the synthesis of the polymerization. The removal of oxygen bubbles from prepolymerized solution is also important before starting the polymerization process because the presence of oxygen molecules may lead to the instability of the final product. The oxygen removal is usually performed by keeping the prepolymerization complex in an inert gas atmosphere such as nitrogen or argon.

Solvents: The solvent is also one of the most important elements for MIP synthesis. It works not only as a porogen for polymerization but also as a rebinding medium during the template/analyte recognition. Although these are used for dissolving the MIP elements, they have further significant roles such as providing porous nature to the polymer matrix which results in the easy mass transfer of the analyte to the recognition sites (Gupta et al. 2016). The further role of the solvent is to provide the homogeneous temperature by dispersing the heat generated during polymerization. Otherwise, in the case of heterogeneous heat, the unwanted reactions may occur. The adsorption nature of the MIP is also affected by the polar nature of the solvent. In the case of noncovalent molecular imprinting, mostly nonpolar solvents are used as these do not affect the interaction between the monomer and template. For example, chloroform is the largely used solvent because it can dissolve many monomers and templates without affecting the interactions between them.

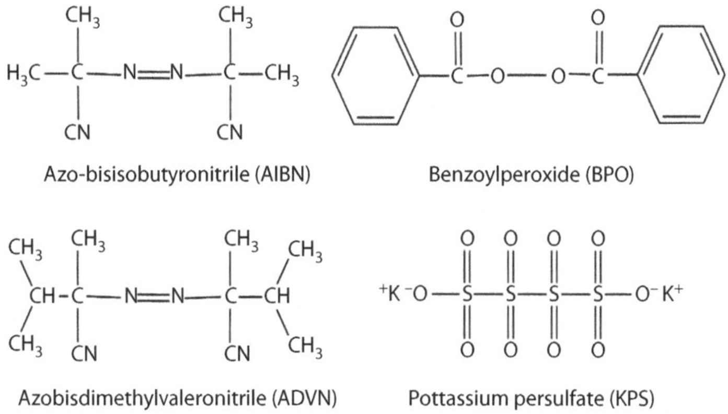

FIGURE 6.2
Few common polymerization initiators used for MIP synthesis.

Polymerization temperature and pressure: The ambient temperature and pressure are also the important parameters which affect the MIP properties. Prior to polymerization, it is necessary to find an equilibrium condition between the template, monomers, and their complexes which can be achieved at a suitable temperature and pressure (O'Shannessy et al. 1989). As the electrostatic interactions are stronger at low-ambient temperature, MIP based on these requires low-polymerization temperature. It has been found that the selectivity of MIP can be improved by polymerization at high pressure (1000 mbar) (Sellergren et al. 1997).

So far, we have discussed the elements and their role for MIP synthesis with their effects on the properties of MIPs. Now, in the next, we shall discuss various procedures for MIP synthesis.

6.1.2 Synthesis Protocols

The selection of a suitable procedure for the preparation of MIP with required properties is also very crucial. Mostly there are two main mechanisms which are followed to produce MIP: free radical polymerization and sol-gel polymerization. The free radical polymerization mechanism is used in several imprinting methods like bulk polymerization, suspension polymerization, emulsion polymerization, and seed polymerization. The first reported and simplest method for the polymerization is the bulk polymerization method. In this method, first the polymerization is performed and after that the resultant product is mechanically crushed or grinded for the preparation of small particles of MIP. The particles prepared are of sizes up to the millimeter. Although the bulk polymerization is the simplest method because it does not require any sophisticated instrumentation, the method suffers from less homogeneous particle size distribution and large template requirement (Wulff 1995). This problem can be overcome by adopting the suspension polymerization method. In this method, mostly used solvents are mineral oils, perfluorocarbon liquid, and water. Among these, perfluorocarbons are preferred because the other solvents may exhibit some adverse effects on the template–monomer complex. There is a broad range of the size of MIP particles varying from micrometers to millimeters depending on the stirring speed and surfactant amount. The method possesses the disadvantages of large size distribution and poor recognition capability due to the influence of solvents on the monomer–template interactions (Wang et al. 2011). The emulsion polymerization method successfully provides MIP particles with homogeneous size distribution and high yield in both water-in-oil and oil-in-water approaches. However, the use of surfactants limits its application in the biomedical industry (Dvorakova et al. 2010). Further, seed polymerization method results in MIP particles with comparatively controlled diameters with approximately similar shape and size. These particles are mostly used in chromatographical applications. The MIP particles prepared using this method show less selectivity due to the use of more liquid suspensions. In addition, the seed polymerization (*in situ* polymerization method) method suffers from the disadvantage of higher-time consumption (Wulff et al. 1977). The particles with submicron range (0.3–10 μm) can be prepared by precipitation polymerization method (Ramstroem et al. 1993). In this method, more than 95% (% wt) of solvent is used as compared to bulk polymerization method. Due to this, the particles are secured from the coalescence which occurs due to the polymer cross-linking and hence there is no requirement of the extra stabilizer. The use of a large amount of solvent and the requirement of strictly controlled ambient conditions are the few restrictions of this method. Further, in the sol-gel method, usually, tetra-alkoxysilane precursors are used. These precursors are initially hydrolyzed to create the colloidal solution (as a sol) which is further polycondensed to the construction of the highly cross-linked silica materials (as a gel). The method possesses several beneficial properties such as easy fabrication steps at room temperature without any thermal decomposition, the compatibility with eco-friendly solvents such as water and ethanol (while in the case of radical polymerization, other solvents such as chloroform, toluene, etc. are used). The pros and cons of the abovementioned methods have been summarized in Table 6.4.

Although aforementioned traditional methods are sufficient to produce the effective MIPs, the resultant still suffers from various problems such as template leakage, irregular material structure, lower-binding ability, and liquid medium incompatibility. To overcome these problems, advanced imprinting strategies are introduced. These have been discussed below.

Surface imprinting technology: Surface imprinting technology is used for the MIP materials where the templates are polymerized at the surface or the proximity of the NIP. The recognition sites are created mostly at the surface or nearby of the MIP which provides good accessibility to the guest

TABLE 6.4

Pros and Cons of Different MIP Synthesis Methods

Polymerization Method	Pros	Cons
Bulk	Easy process, simple instrumentation, less contamination	Time-consuming, mixed shape and size, lower affinity of binding sites
Suspension	Simple and one-step polymerization, spherical-shaped particles	Less binding affinity, large-sized particles
Emulsion	Monodispersed, water-soluble particles with high yield	Use of surfactant, less imprint preparation
Seed	Monodispersed spherical particles in a controlled manner, can be used in HPLC	Time-consuming, more laborious work required
Precipitation	One-step synthesis, spherical particles with uniform shape and size, highly stable	High-dilution factor, large amount of template required
Sol-gel	Eco-friendly solvents, easy fabrication steps at room temperature	Lack of monomer

template molecule for recognition (Chen et al. 2011). This type of imprinting is mostly used for the templates with large sizes such as cells, viruses, and proteins because the large size hinders the imprinted sites usually while removing/rebinding in traditional MIPs. However, the surface imprinting suffers from limited imprinted sites due to the finite surface area of the substrate. The number of binding sites can be enhanced by using the substrate having large-surface area. Compared to the bulk imprinting, the surface imprinting shows high-binding capacity, excellent selectivity, better reproducibility, and fast adsorption kinetics (Chen et al. 2016).

Multitemplate imprinting technology: Usually, the synthesis steps of MIP include the use of a single template. However, the use of single template inhibits the applications of MIP for the simultaneous recognition of more than one template. The fabrication process for the synthesis of MIPs having multiple types of imprinted sites is called as multitemplate-imprinting technology. In this method, more than one template is used in the preparation of single polymer matrix. These imprinted polymers can be used in separation, extraction, and detection of more than one analyte simultaneously (Chen et al. 2016).

Multimonomer-imprinting technology: This is mostly used when noncovalent imprinting is performed. As in the case of noncovalent imprinting, the interactions between monomer and templates are noncovalent in nature which can be enhanced by multipoint interactions (Athikomrattanakul et al. 2011). Thus, in multimonomer-imprinting technology, two or more functional monomers are utilized for the noncovalent interaction with the template. The use of two or more functional monomers results in an enhancement in selective nature of MIP. This is also mostly used for the creation of macromolecules imprints. However, this method is not easy and the selection of two complementary monomers which interact with templates noncovalently is an important task. There is also another important method for the synthesis of MIPs, which is called as nanoimprinting. In this method, MIP nanoparticles are prepared which enhance the various properties of MIP. These have been found to be more effective in various applications. The evolution and importance of these have been further discussed in Section 6.2.3.

MIP characterizations: After the synthesis of MIP, its characterization is very important which gives the information of particle size, stability, porosity, binding/unbinding kinetics, etc. This information is used to decide the application of synthesized MIP. The information regarding the surface morphology of MIP is generally evaluated from scanning electron microscopy (SEM) and transmission electron microscopy (TEM) analyses. The topographical information regarding the smoothness of MIP thin film is analyzed by atomic force microscopy (AFM) and fluorescence technique. The information regarding the surface interactions and porosity can be investigated by the Brunauer–Emmett–Teller (BET) analysis, while the thermal stability is measured by thermogravimetric analysis (TGA). Further, the binding properties of MIP such as template–monomer interaction and total removal of the template from NIP can be observed through the Fourier transform infrared

(FTIR) spectroscopy, ultraviolet-visible (UV–Vis) spectroscopy, and nuclear magnetic resonance (NMR). There are various other characterization methods such as x-ray photoelectron spectroscopy (XPS) and vibration sampling magnetometry (VSM) which are used to find the information regarding the template–monomer interactions and magnetic properties, respectively (Chen et al. 2016).

6.2 Types of Molecular Imprinting

As discussed in Section 6.1.1, MIPs are classified into two types depending upon the nature of the interaction between monomer and template. These are covalent imprinting and noncovalent imprinting. In the following sections, we shall provide a brief discussion of these.

6.2.1 Covalent Molecular Imprinting

The covalent imprinting refers to the MIPs which are prepared by the covalent interactions between the template and monomer (Wulff et al. 1977). In this method, initially template–monomer complex is prepared by covalent linkage and then the polymerization process is performed. Various pairs of chemical conjugates such as amine-aldehyde, acid-amine, and diol-ketones show the covalent interactions. Hence, the covalent imprinting can be performed using one of these pairs as the template–monomer conjugates. Based on these, the imprinting of Shiff's base, esters, and ketals/acetals has been performed. The covalent imprinting has both advantages and disadvantages which have been discussed below.

Advantages: (a) The imprinted sites prepared by covalent imprinting, defined as the template–monomer complexes, are quite stable and stoichiometric, and (b) these polymers are highly stable and can be retained in various harsh environments such as high/low temperature, high-polar solvents, high/low pH value, etc. This is because of the covalent interactions between template and monomer.

Disadvantages: (a) The synthesis of the monomer–template complex is quite expensive and requires certain strict environmental conditions, (b) there are limited numbers of template–monomer complexes with reversible covalent interactions, and (c) the kinetics of template binding/rebinding is slow.

An example of covalent imprinting of sialic acid is shown in Figure 6.3 (Kugimiya et al. 1995).

In this process, sialic acid is the template molecule which is an ester. 4-Vinylphenylboronic acid works as monomer, while EGDMA is the cross-linking agent. The polymerization of sialic acid–4-vinylphenylboronic acid complex is performed under UV light at 4°C. Finally, the template removal is performed by treating the polymer with hydrochloric acid and methanol.

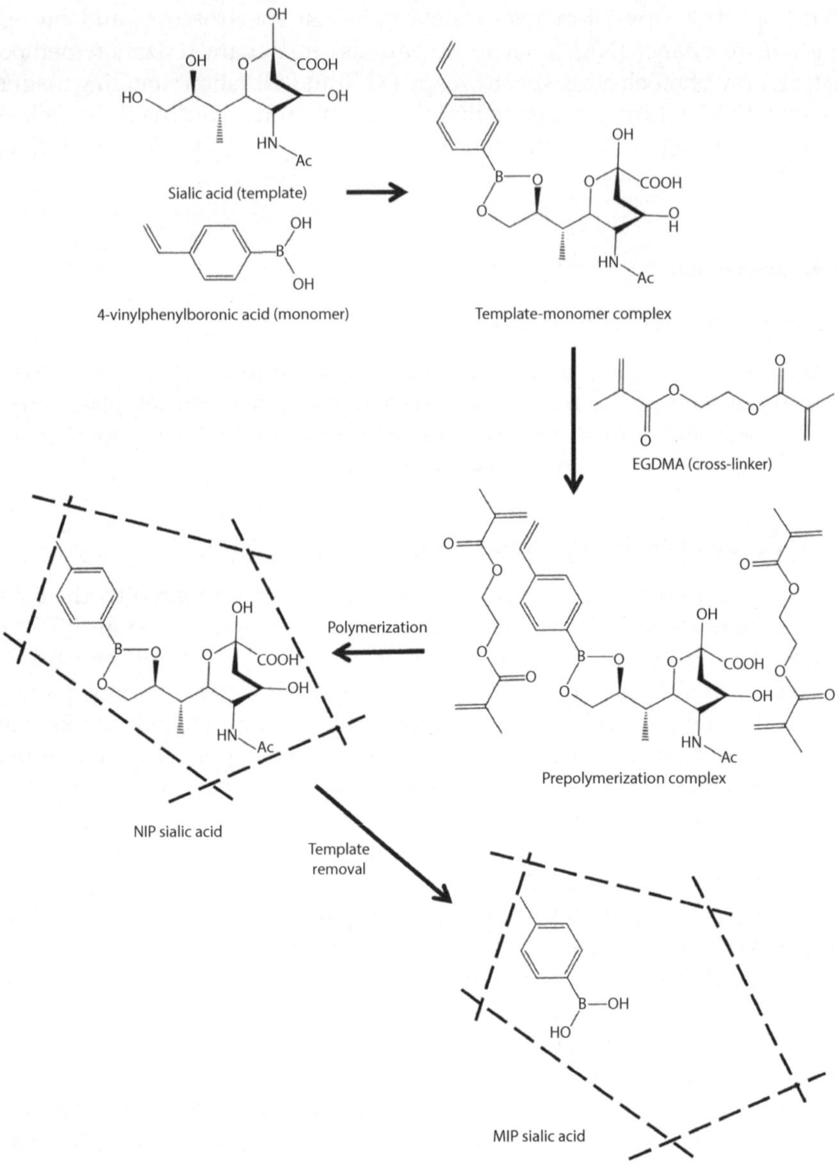

FIGURE 6.3
Mechanism of covalent imprinting where sialic acid is used as a template.

6.2.2 Noncovalent Molecular Imprinting

This type of imprinting was invented by Mosbach and his coworkers in 1985 (Whitcombe et al. 1995). They showed that the preparation of imprinted sites within a polymer matrix does not require any covalent linkage. In this type of imprinting, the interaction between the template and monomer is noncovalent in nature such as π–π, ionic, electrostatic, hydrogen bonding, and Van der Wall forces. The synthesis procedure for these types of MIPs is very simple as compared to covalent imprinting. The preparation steps involve simple mixing of all the agents, which provides a prepolymerized complex. The complex is then polymerized and the removal of template is performed for the creation of imprinted sites. Mostly the noncovalent-imprinted polymers are prepared using hydrogen bonding between the template and monomer. This is achieved by using amines as template and MAA as the monomer (Schirhagl 2013). Like covalent imprinting, noncovalent imprinting also has several advantages and disadvantages as discussed below.

Advantages: (a) The method provides the freedom from the fabrication of covalent template–monomer complex, (b) the removal of the template from NIP is very simple and can be possible in mild ambient conditions as templates are bound with the weak interactions, and (c) due to the noncovalent interaction, the binding/removal kinetics is fast.

Disadvantages: (a) The shape and size of imprinted sites are less defined as compared to covalent imprinting, (b) the ambient conditions for polymerization should be chosen carefully, and (c) as compared to covalent imprinting, the imprinted sites are less selective.

Figure 6.4 shows an example of the noncovalent imprinting, where atrazine is a template molecule. In this example, MAA is used as the functional monomer. The hydrogen bonding occurs due to the presence of the carboxylic group (–COOH) in MAA. EGDMA works as the cross-linker, while AIBN's role is of an initiator.

Thus, both types of imprintings have their own pros and cons which can be reduced by a new type of imprinting which has the combined nature of covalent and noncovalent imprintings, known as semicovalent imprinting (Chen et al. 2016). The method offers the covalent binding between the template and monomer, while after removal the rebinding of the guest template is because of noncovalent interaction. This method possesses advantages of high stability due to covalent interaction, while the fast rebinding/unbinding is because of the noncovalent interaction.

6.2.3 Molecular Imprinting in Nanostructures

Although the MIPs have several advantages such as high selectivity, stability, and fast recognition, these properties can further be improved by the combination of molecular-imprinting technology and nanotechnology, which causes the preparation of MIPs up to nanoscale (Wackerlig and Lieberzeit 2015). The improvement is because of the better site accessibility and more binding capacity for the template rebinding. Different types of nanostructures such as nanoparticle, nanoflower, nanowire, and nanotube of MIP can be prepared by simply varying the synthesis procedure of traditional MIP (Wackerlig and Lieberzeit 2015). The particles prepared by traditional

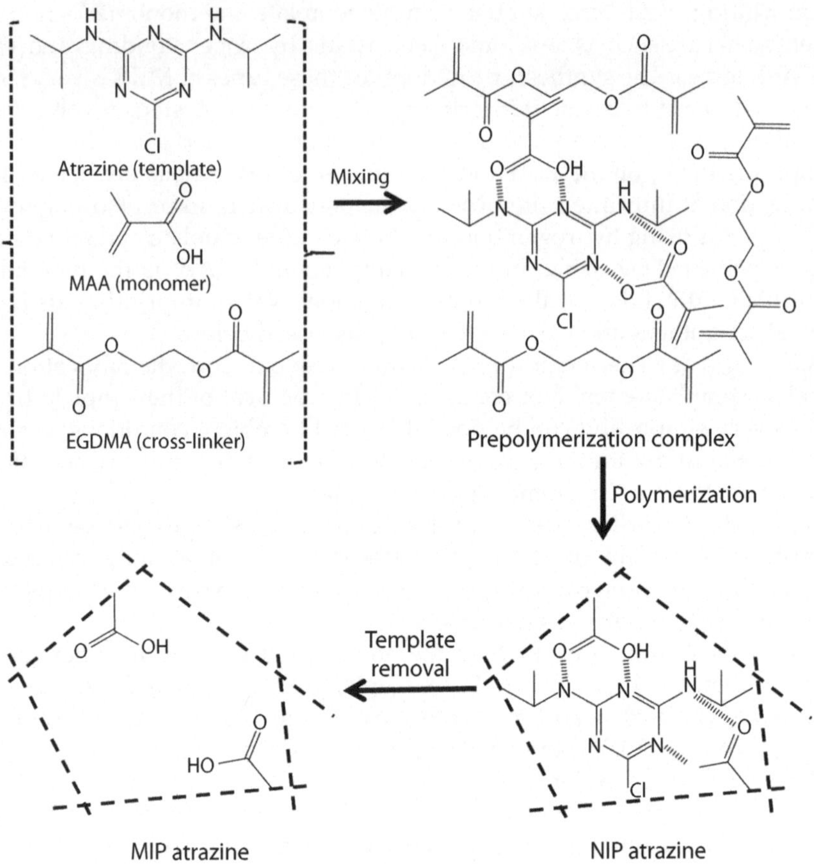

FIGURE 6.4
Mechanism of a noncovalent imprinting where atrazine is used as a template.

methods are in the dimension of micrometers and usually prepared by grinding. This process leads to the MIP product with irregular shape and size; thus, the synthesized particles inherently show low yield. Due to this, the MIP particles are found to be less suitable for industrial applications. This is the main reason for the emergence of nanotechnology in molecular imprinting. MIP nanoparticles are usually prepared by varying the few parameters in traditional methods, discussed in Section 6.1.2. These parameters are the type of solvent, value of M:T ratio, cross-linking degree (usually 60%–90%), ambient temperature, etc. Further as compared to traditional MIP, nano-MIP has following other advantages: (a) homogenous binding affinity for all the imprinted sites, (b) increase of particle solubility in aqueous medium due to small-particle size, (c) more controlled synthesis procedure, (d) increased number of binding sites due to high surface to volume ratio, (e) increase in binding/unbinding kinetics, and (f) ease of template trace removal. Thus, considering these, the nanostructured MIPs are making evolutions in the field of molecular-imprinting technology and increasing its prospects in various *in vivo* biological applications.

6.3 Molecular-Imprinting Polymer as a Floor for Sensing

Till now, we have discussed the fundamentals, types, and various approaches for the synthesis of MIPs. In this section, we shall discuss the role of MIPs in various applications especially in sensing. As mentioned earlier, an effective MIP has several advantageous properties such as highly specific template recognition, high-adsorption ability, fast binding–unbinding kinetics, homogeneous particle size, good imprinting efficiency, and much more. These properties make MIP play an effective role in various applications of extraction, separation, purification, assay formation, and sensing. These roles are summarized in Figure 6.5.

Since MIPs are known for the preparation of highly specific recognition sites of the template molecule, this makes MIP to play a key role in sensing technology where high selectivity is essential for a good sensing device. The MIP, apart from having receptor sites for the detection of template molecule, is also used as the sensing layer for the template detection. When template comes near the imprinted sites on MIP, it binds covalently/noncovalently (depending upon the type of imprinting) with the imprinted site and changes the chemical, electrical, or optical properties of the MIP matrix. In the case of optical sensor, the change in optical properties (such as refractive index) is detected using various transducing methods such as fluorescence, chemiluminescence, colorimetric, SPR, LSPR, LMR, SERS, etc. As mentioned earlier, the sensors based on MIPs are highly selective as the recognition

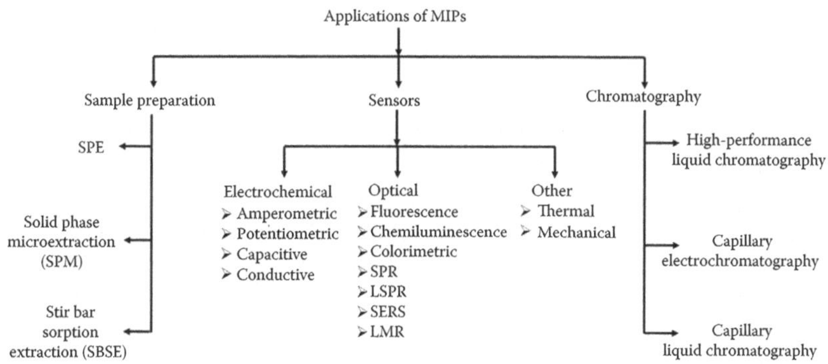

FIGURE 6.5
Structural representation of MIP's applications.

sites in MIPs are 3D cavity-like structures having the complementary shape and size of the template to be detected. Any molecule other than template if comes near these imprinted sites, due to the different structure, it does not fit within these sites; thus, no interaction happens between guest molecule and the imprinted sites. Hence, no change in the properties of MIP matrix is observed while interacting with any interfering molecule other than the template. This is the reason why molecular-imprinting technology is so promising in sensing applications. In addition, MIP is an artificially prepared recognition medium; it has several other advantages over the natural receptor.

In the present era, usually, sensors for biomedical applications, food analysis, and environmental monitoring have been fabricated using naturally occurring receptors such as enzymes, antigens, antibodies, etc. Although these are broadly used, they suffer from many disadvantages also. The main disadvantageous factors are poor chemical stability, high cost, complex and expensive preparation protocols, and sophisticated immobilization process. All these drawbacks of natural receptors can be overcome by introducing the tailor-made MIP receptors. MIP possesses numerous unique properties like high stability even at the harsh environment, low cost, easy immobilization ability, quick binding/unbinding kinetics, fast response, and much more. By considering these beneficial properties, MIPs are showing their great potential in chemical and biological sensing technology and are replacing antigen-antibody and enzyme-based sensors due to better sensing parameters of MIP-based sensors such as high sensitivity and selectivity, fast response, better stability, and low cost. In the next section, we shall discuss some of the MIP-based optical sensors for applications in food safety, environment monitoring, and biomedical sciences.

6.4 Applications

6.4.1 Food Safety

Nowadays, the food contaminants are causing various hazarding effects to the human and animals. Thus, the monitoring of the contaminants in food products or food safety is essential for humans and environment. A number of detection methods have been reported for the detection of numerous adulterations in food stuffs using molecular-imprinting method. Few of these have been discussed below.

Melamine: Melamine is a nitrogen-rich compound ($C_3H_6N_6$) and due to excess of nitrogen within its chemical structure, it works as the flame-retarding compound. It is also used in the fabrication of fabrics, adhesives, fertilizers, plastics, paints, and kitchenware. However, due to the excess of nitrogen, it is used as an adulterant in milk and various foodstuffs to increase the protein content which is calculated from the nitrogen content within the sample. Its excess beyond a certain concentration is harmful to the human and may result in reproductive system damage and cancer. Thus, the sensing of melamine in foodstuffs is important. Optical sensors for the detection of melamine using molecular imprinting have been reported. As an example, we discuss a fiber-optic surface plasmon resonance-based melamine sensor fabricated using MIP layer as the sensing layer (Shrivastav et al. 2015a). The sensing probe was fabricated over 1-cm-unclad length in the middle portion of the 15-cm-long plastic clad silica fiber. The unclad portion was coated with a 40-nm-thick silver film followed by the coating of a MIP layer having melamine as template (MIP melamine). MIP melamine was prepared using MAA as the monomer, EGDMA as cross-linker, AIBN as reaction initiator, and benzene as the porogen. The fabricated probe was characterized using the experimental setup shown in Figure 4.18 in Chapter 4. Figure 6.6a shows the SPR spectra for varying melamine concentration from 10^{-7} to 10^{-1} M. As can be seen, the SPR spectrum shifts as the concentration of melamine increases. The shift in resonance wavelength for the change in concentration from 10^{-7} to 10^{-1} M was observed as 19.52 nm. Further, the sensor possesses the maximum sensitivity of 10.1 nm/log M near melamine concentration of 10^{-7} M and the limit of detection as 9.87×10^{-9} M. The study showed that MIP prepared with template concentration of 0.2 mM possesses the best performance, as it is an important parameter affecting the MIP properties. The sensor also showed highly selective nature toward melamine detection as shown in Figure 6.6b.

The sensitivity of the probe was further enhanced by including a 3-nm-thick silicon (Si) film between silver and MIP layer (Shrivastav and Gupta

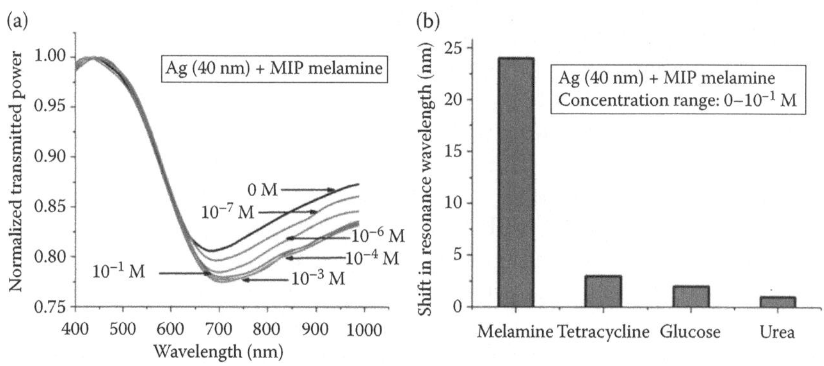

FIGURE 6.6
(a) Spectral response (b) selectivity of a fiber-optic SPR sensor for melamine detection with concentrations from 0 M to 10^{-1} M. (Reprinted from *Sens. Actuators B* 212, Shrivastav, A. M. et al., Fiber optic SPR sensor for the detection of melamine using molecular imprinting, 404–410. Copyright 2015, with permission from Elsevier.)

2016). The silicon film, being of high-refractive index, causes the enhancement of electric field intensity at Si/MIP interface. The sensor showed the sensitivity of 76.0 nm/log M at the melamine concentration of 10^{-8} M which is much better than that obtained without silicon layer. Further, the sensor with modified probe possessed the value of LOD as 4.3×10^{-11} M which is much lower than that obtained in the case of sensor without silicon layer. A colorimetric melamine sensor has also been reported that uses MIP magnetic nanoparticles (You et al. 2016). In this, MIP nanoparticles were synthesized using the mini-emulsion method and the reagents used were MAA (monomer), acrylamide (AM/monomer), EGDMA (cross-linker), chloroform (solvent), and sodium dodecyl sulfate (SDS/surfactant). The binding of melamine with the imprinted sites of MIP causes the increase in the effective refractive index of MIP nanoparticles, and hence a redshift in absorption spectrum was found for the increase in the concentration of melamine sample around nanoparticles. The sensing method showed the redshift of peak absorbance wavelength of about 200 nm for melamine concentration change from 0 to 0.1 mg/mL, while the sensor possessed the detection limit of 10^{-5} M.

Tetracycline: It is a type of antibiotic which is used for the treatment of various bacterial infections like acne, skin infections, and pneumonia. Its excess amount in various foodstuffs results in the allergic effects on human body and animals. Thus, the detection of tetracycline residue concentration is important for food safety. A fiber-optic SPR sensor using MIP tetracycline for the determination of tetracycline concentration was reported (Verma and Gupta 2013a). The sensing probe was fabricated in the same manner as the SPR-based melamine sensor discussed above (Shrivastav et al. 2015a). The synthesis steps of MIP tetracycline involve mixing of monomers acrylamide: bis-acrylamide (AM:BIS) with template tetracycline in deionized water, named as the master solution. The master solution was then mixed with phosphate buffer, acrylic acid (AA), ammonium persulfate (APS), and tetramethylethylenediamine (TEMED) resulting in the prepolymerized solution. The solution was then kept for polymerization at 60°C for 4 h and the NIP was prepared. Finally, the imprinted sites were created by dipping the NIP-coated fiber-optic probe in SDS/acetic acid solution for 2 h. The prepared probe was characterized using the samples of tetracycline in the concentration range of 0–0.96 μM. The SPR spectra of the sensor obtained for different concentrations are shown in Figure 6.7a. The limit of detection of the sensing probe obtained using the sensitivity of the sensor is 1 nM. This work was further extended by adding a silver nanoparticle (20–30 nm) layer between silver and MIP tetracycline layers (Shrivastav et al. 2015b). The performance of the sensor got improved in terms of operating range, sensitivity, and detection limit. This is because of the combined transducing phenomenon of SPR and LSPR. The absorbance spectra of the sensing probe for the tetracycline concentration in the range of $0–10^{-4}$ are shown in Figure 6.7b. The reported LOD value of the sensor is 2.2×10^{-9} M.

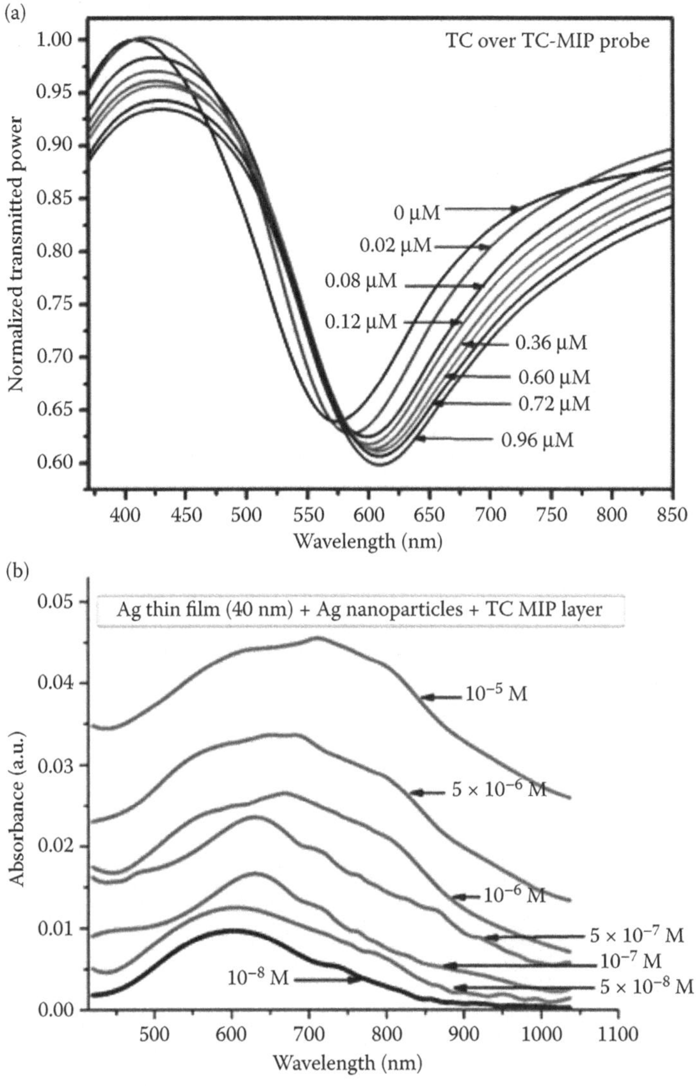

FIGURE 6.7
Response of (a) SPR and (b) LSPR + SPR-based fiber-optic tetracycline sensors utilizing molecular-imprinted polymers as recognition medium. ([a] Verma, R. and B. D. Gupta. 2013. Optical fiber sensor for the detection of tetracycline using surface plasmon resonance and molecular imprinting. *Analyst* 138, 7254–7263. Reproduced by permission of The Royal Society of Chemistry. [b] Reprinted from Shrivastav, A. M. et al. 2015. Localized and propagating surface plasmon resonance based fiber optic sensor for the detection of tetracycline using molecular imprinting. *Mat. Res. Exp.* 2, 035007. With permission from IOP Science.)

Erythromycin: Erythromycin is a type of antibiotics and it is used for the treatment of several bacterial diseases. Naturally, erythromycin is absent from the foodstuffs such as honey and milk but its heavy use in apiculture and cattle industries causes the erythromycin residues in milk, honey, and the concerned foodstuffs. Its high concentration is harmful to the human body. Therefore, its sensing is also essential for food safety. An erythromycin sensor using molecular-imprinted erythromycin nanoparticles was reported using gold SPR chip (Sari et al. 2016). The molecular-imprinted nanoparticles were synthesized using two-phase mini-emulsion method where erythromycin and MAA were used as template and monomer, respectively. Molecular-imprinted nanoparticles were coated over gold-coated glass chip and the probe was characterized using a commercial SPR system. The sensor chip was characterized for 0–50 ppm concentration range of erythromycin solution. The method showed the detection limit of 0.29 ppm (0.4 μM). Later, another SPR-based erythromycin sensor was reported using MIP erythromycin nanoparticles over an optical fiber substrate (Shrivastav et al. 2017). The molecular-imprinted erythromycin nanoparticles were coated over silver-coated unclad core of the fiber for the probe fabrication. The operating range of the sensor was 0–100 μM. The sensing mechanism and its response for varying erythromycin concentrations are shown in Figure 6.8. The sensor was reported to possess LOD and LOQ values of 1.62×10^{-3} and 6.14×10^{-3} μM, respectively. Further, the method also showed its applicability in real milk and honey samples.

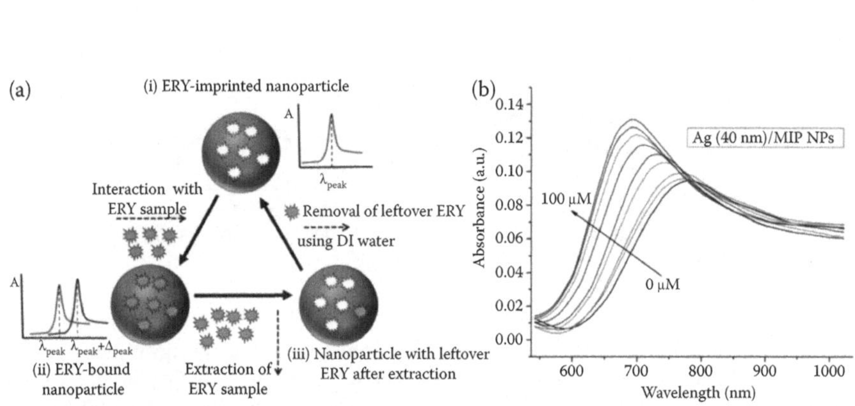

FIGURE 6.8
(a) Pictorial representation of the sensing mechanism, and (b) absorbance spectra of fiber-optic erythromycin sensor fabricated using SPR and molecularly imprinted nanoparticles with varying erythromycin concentration from 0 to 100 μM. (Reprinted from *Biosens. Bioelectron.* 90, Shrivastav, A. M. et al., Highly sensitive and selective erythromycin nanosensor employing fiber optic SPR/ERY imprinted nanostructure: Application in milk and honey, 516–524. Copyright 2017, with permission from Elsevier.)

6.4.2 Environmental Monitoring

The environmental pollution is always a serious topic of concern for the researchers. It is continuously increasing due to various contaminants such as pesticide and antibiotic residues, toxic metal ions, hazardous gases, etc. Thus, it is essential to detect these in the environment and water for the safety of living species. In this section, we shall provide a brief discussion of some of the reported studies on the optical methods for the detection of various pesticides, insecticides, metal ions, and antibiotics using molecular-imprinting technique.

Profenofos: Profenofos is a pesticide which belongs to the organophosphorus pesticides (OPPs) family, which is broadly used for the controlling of the pest on the vegetables, rice, cotton, etc. The presence of OPPs in human body can result in injurious effects on the nervous system even at low concentrations. The extensive use of profenofos may lead its residues in water, hence the detection of profenofos is important. An optical method for the detection of profenofos was reported using SPR over a gold-coated glass chip (Dong et al. 2012). In this method, immobilization of MIP profenofos layer was performed over the gold-coated glass chip through a self-assembled monolayer (SAM). The thickness of MIP layer was about 20 nm. The sensor's performance was checked from the characterization of the probe with profenofos-spiked samples in tap water having the concentration range of 0.001–0.1 μg/mL. The sensor possessed high sensitivity and selectivity, good stability and reproducibility with the detection limit of 3.6×10^{-4} mg/mL. Later, a fiber-optic sensor for the detection of profenofos was reported using SPR and molecular-imprinting technique (Shrivastav et al. 2016a). In this method, the probe was fabricated using the dip coating of MIP profenofos layer over 40-nm-thick silver coated unclad core of a multimode fiber. The free radical polymerization was used for the synthesis of MIP profenofos. The various elements used for the fabrication of the probe such as template, monomer, cross-linker, and solvents are profenofos, MAA, trimethylolpropane trimethacrylate (TRIM), and dimethyl sulfoxide (DMSO), respectively. The sensor's performance was checked for profenofos concentration in the range of $0–10^{-1}$ M. Figure 6.9 shows the spectral response, calibration, and sensitivity curves of the reported method. The method also showed its potential for the determination of profenofos in wastewater samples.

Atrazine: Atrazine is basically an herbicide which is mostly used for controlling the broad-leaf weeds in various crops such as grapes, corn, sugarcane, pine, etc. It belongs to the Restricted Use Pesticide (RUP) category which means that it can only be used by registered professionals, while its use is restricted for public. It is generally found in ground and drinking water samples. It is highly toxic and affects the ecosystem and human being with various diseases such as cancer, damage, reproductive system, hormone disruption, and immunosuppression. These reasons make its sensing important. The sensing of atrazine using molecular-imprinting technique

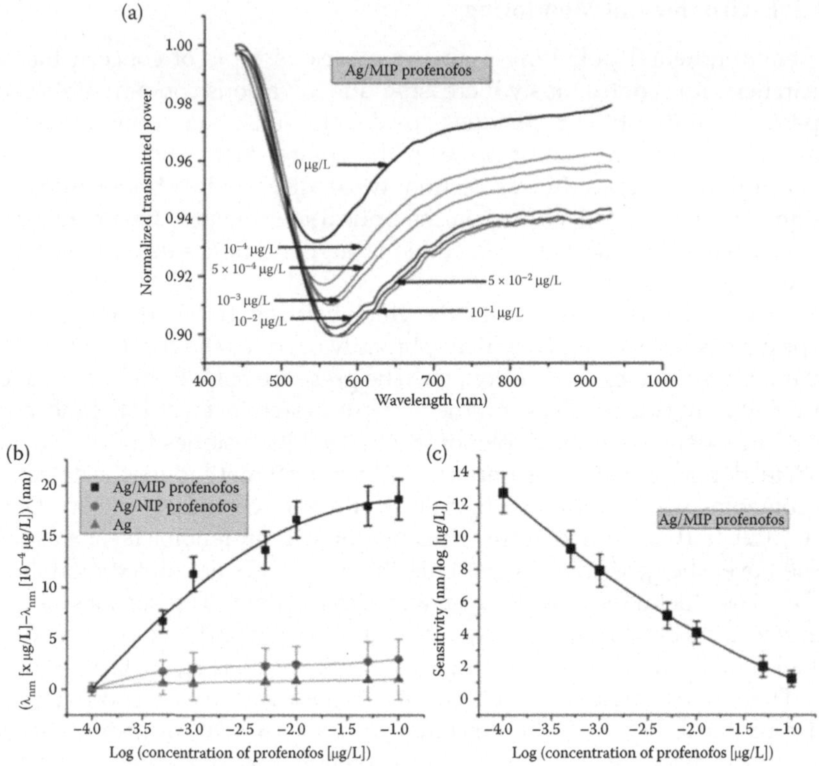

FIGURE 6.9
(a) SPR response, (b) calibration curve, and (c) sensitivity of the fiber-optic probe for different profenofos concentration solutions in the range from 0 to 10^{-1} µg/L. The sensing is based on SPR and molecular-imprinting technique. (Reprinted from *Biosens. Bioelectron.* 90, Shrivastav, A. M. et al., Highly sensitive and selective erythromycin nanosensor employing fiber optic SPR/ERY imprinted nanostructure: Application in milk and honey, 516–524. Copyright 2017, with permission from Elsevier.)

over fiber-optic SPR platform was reported (Agrawal et al. 2016). The process of probe fabrication was similar to the profenofos sensor discussed above with a difference of sensing layer. For the profenofos sensor, MIP profenofos was used as the sensing layer, while in this case MIP atrazine was used as the sensing layer. The sensor possessed an operating range of 0–10^{-7} M with detection and quantification limits of 1.92×10^{-14} and 7.61×10^{-14} M, respectively.

Metal ions: The detection of metal ion concentration is always a topic of interest for the researchers due to their adverse effects on human health. The detection of various metal ions such as Cu(II), Pb(II), Hg(II), As(IV), and many more has been reported. We shall discuss few methods reported for the metal ion detection using molecular imprinting. A highly sensitive and selective sensing method for the detection of Cu(II) ions using ion-imprinted nanoparticles was reported (Shamsipur et al. 2010). For the fabrication of the probe, Cu-imprinted nanoparticles were prepared in two steps using precipitation polymerization method. The first step involved the preparation of metal–ligand complex by mixing the 1-hydroxy-4-(prop-2-enyloxy)-9,10-anthraquinone (AQ, ligand) and $Cu(NO_3)_2 \cdot 3H_2O$ (template ion) in acetonitrile solvent. The second step was followed by adding EGDMA (cross-linker) and AIBN (initiator) in the metal–ligand complex to cause the formation of prepolymerization complex. The thermal polymerization of resulted solution was performed by keeping the solution at 60°C for 24 h. After polymerization, the solution was having the nanoimprinted nanoparticles in the precipitation form. The particles were then washed by methanol/water solution for the creation of Cu(II) ion imprints. The particles were characterized by UV–Vis spectroscopy using Cu(II) solutions of varying concentrations. The change in absorbance was found for varying concentrations due to the change in the effective refractive index of the imprinted particles, which was due to the binding of Cu(II) ions to the corresponding imprints. The method gave the relative standard deviation of 2.6% with LOD of 0.1 ng/mL.

6.4.3 Biomedical Uses

Molecular-imprinting-based sensors are also showing their potentials in the detection of the analytes required in the biomedical applications. The monitoring of vitamins, glucose, urea, etc. is essential in the biomedical industry and numerous sensors have been reported for their detection using molecular imprinting technology. In this section, we shall, briefly, discuss some of these.

Vitamin B_3 (3-Pyridinecarboxamide): Vitamin B_3 is one of the vitamins which belong to vitamin B group. It is used for maintaining the proper metabolism, breathing, and healthy skin. Increase in its concentration in human body may result in various hazard effects such as wheezing, diarrhea, skin problems, etc. In view of this, an SPR and molecular-imprinted hydrogel-based optical fiber sensor for the detection of vitamin B_3 was reported (Verma and Gupta 2013b). The MIP was synthesized using vitamin B_3 as template molecule, while AM/BIS, TEMED, and APS as monomer, cross-linker, and initiator, respectively. The sensing probe was fabricated by coating a MIP layer having vitamin B_3 as template over the silver-coated unclad core of the fiber. The characterization of sensing probe was performed using aqueous solutions of vitamin B_3 having concentrations in the range of 0–10 mg/mL. A change of 16.651 nm in resonance wavelength was observed for the concentration change from 0 to 10 mg/mL.

Vitamin C: Vitamin C, also known as ascorbic acid (AA), is present in fruits, vegetables, soft drinks, and pharmaceuticals. Its application in our body is to protect from oxidative stress and is used for the treatment of Parkinson's disease and cancer. However, its high concentration can cause harmful effects such as gastric irritation and renal problems. A fiber-optic sensor for the detection of ascorbic acid has been reported using SPR technique and ascorbic-acid-imprinting polyaniline (PANI) film (Shrivastav et al. 2015c). MIP film was prepared by dissolving AA in PANI solution followed by its extraction. Dissolving AA in PANI results in the freezing of AA molecules in PANI matrix, while the extraction of AA causes the creation of imprinted sites within the PANI matrix. MIP layer was coated over the silver-coated unclad core of the fiber. A schematic of the sensor probe is shown in Figure 6.10a. When AA molecules come near the domains of AA-imprinted sites within the MIP PANI matrix, they interact noncovalently and change the effective refractive index of MIP layer.

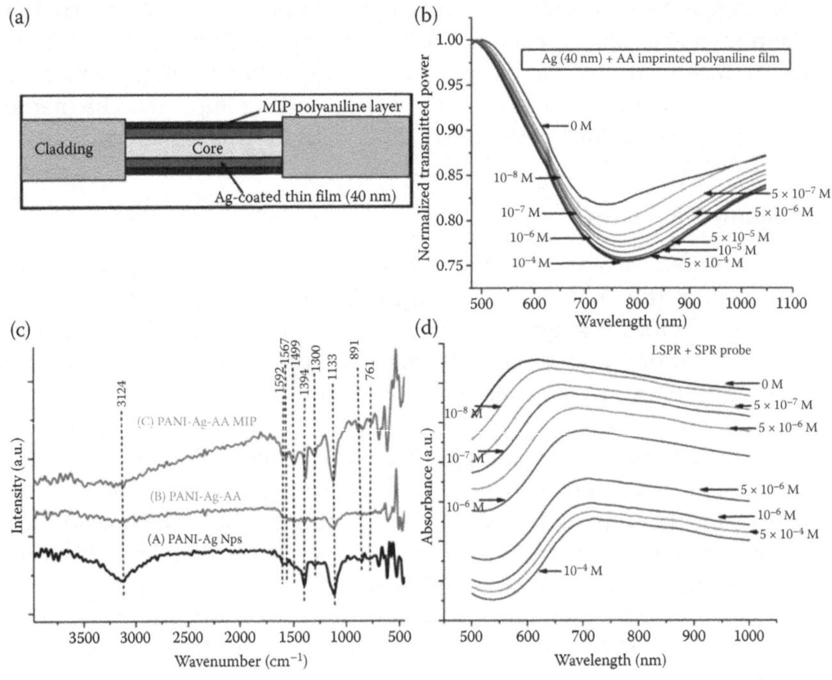

FIGURE 6.10
(a) Schematic, and (b) transmitted spectra of fiber-optic SPR sensor probe for ascorbic acid detection using Ag/molecularly imprinted PANI layer, (c) FTIR spectra of (A) PANI–Ag nanocomposite, (B) PANI–Ag–NIP ascorbic acid nanocomposite, and (C) PANI–Ag–MIP ascorbic acid nanocomposite to ensure the imprinting of ascorbic acid, and (d) absorbance spectra of fiber-optic probe using Ag/PANI–Ag–MIP ascorbic acid nanocomposite realizing both SPR and LSPR. The concentration range of ascorbic acid from 0 to 10^{-4} M. ([a, b] With kind permission from Springer Science+Business Media: *Plasmonics*, Surface plasmon resonance based fiber optic sensor for the detection of ascorbic acid utilizing molecular imprinted polyaniline film, 10, 2015, 1853–1861, Shrivastav, A. M. et al.; [c, d] Reprinted from Shrivastav, A. M. et al. 2016. A localized and propagating SPR, and molecular imprinting based fiber-optic ascorbic acid sensor using an in-situ polymerized polyaniline–Ag nanocomposite. *Nanotechnology* 27: 345501. With permission from IOP Science.)

This change is recognized by the shift in resonance wavelength of SPR spectrum shown in Figure 6.10b. The operating range of the sensor was reported to be from 10^{-8} to 10^{-4} M of ascorbic acid concentration and the shift in resonance wavelength for this change in concentration was found to be 22 nm. The sensor possessed the sensitivity of 26.384×10^{8} nm/M near-zero concentration, while its detection limit was 1.28×10^{-10} M. The performance of the fabricated probe was further improved in an extended work (Shrivastav et al. 2016b). In this study, the sensitivity of the sensor was enhanced by introducing the silver nanoparticles within the PANI matrix. This was done by *in situ* preparation of Ag–PANI composite, and the nanocomposite polymer was used for the preparation of MIP. The probe fabrication process involved optimizations of various parameters such as PANI–Ag ratio, polymerization time, immobilization period, etc. The FTIR response of Ag–PANI, AA–Ag–PANI, and molecularly imprinted AA–Ag–PANI composites is shown in Figure 6.10c. To characterize the probe, absorbance spectra were recorded for varying concentrations of AA. The spectra are shown in Figure 6.10d, and it may be seen that the peak absorbance wavelength increases as the concentration of AA increases. The shift in peak absorbance wavelength was found to be 86 nm for AA concentration change from 10^{-8} to 10^{-4} M, while the enhanced sensitivity was found to be 45.1×10^{8} nm/M near-zero concentration.

Cortisol: It is a stress biomarker, and when any person is having stress, its concentration in saliva increases. Thus, the monitoring of its concentration is important for mental disease diagnostics. A fluorescence-based cortisol sensor using molecular-imprinted core shell nanoparticles was reported (Murase et al. 2016). The cortisol-imprinted core shell molecular nanoparticles were prepared using cortisol-21-monomethacrylate (template), itaconic acid and styrene (monomers), divinylbenzene (cross-linker), and tetrahydrofuran (solvent). Monodispersed poly(styreneco-DVB) particles used as seed particles were the cores for imprinted nanoparticles. The fluorescence of these particles was measured by adding the cortisol solution of different concentrations in the core shell nanoparticle solution. The change in fluorescence was observed due to the interaction of cortisol molecules with the imprinted sites over the nanoparticles. The method showed the detection limit of 80 nM. Later, another study using molecular-imprinting technology was reported for cortisol sensing (Usha et al. 2017). In this method, the imprints of cortisol molecules were prepared in ZnO–PPY (polypyrrole) composite and 12-nm-thick ZnO layer was used as transducing layer which realizes LMR. Figure 6.11a shows a pictorial representation of fabricated probe. The cortisol samples for the characterization of the probe were prepared in artificial saliva. The LMR absorbance spectra for different concentrations of cortisol in the range of $0–10^{-6}$ g/mL are shown in Figure 6.11b, while the calibration curve of the sensor is shown in Figure 6.11c. The fabricated probe showed the detection limit of 29.5 fg/mL which was the lowest detection limit ever reported.

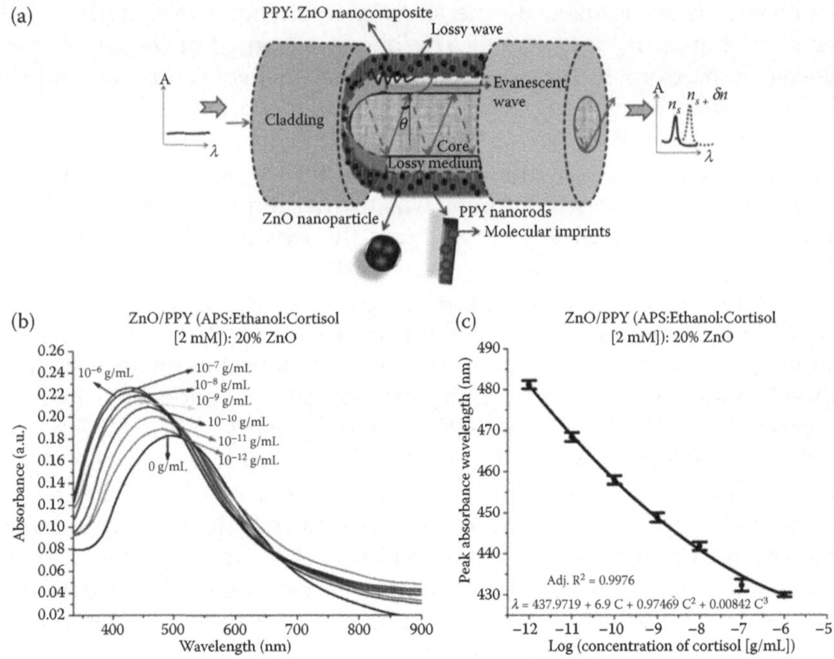

FIGURE 6.11
(a) Schematic, (b) absorbance spectra, and (c) calibration curve of LMR-based fiber-optic cortisol sensor probe using layers of ZnO/ZnO: PPY cortisol-imprinted composite. (Reprinted from *Biosens. Bioelectron.* 87, Usha, S. P. et al., A contemporary approach for design and characterization of fiber-optic-cortisol sensor tailoring LMR and ZnO/PPY molecularly imprinted film, 178–186. Copyright 2017, with permission from Elsevier.)

6.5 Summary

The method of freezing of molecules in a template medium and creating their imprints at their positions is called molecular imprinting. The snapshot of the templates in the polymer matrix works as the highly selective recognition sites for the template molecules. Therefore, MIPs are showing their potential in sensing applications by replacing the naturally occurring receptors such as enzymes, antigens-antibodies, nucleic acids, and DNAs. In addition, natural receptors also suffer from the disadvantages of high cost, low-chemical stability, low-binding kinetic, storage problems, and complex immobilization. In this chapter, we have provided a simple introduction of MIP and its various applications. Further, its fundamentals including synthesis steps and its role in sensing applications have been discussed. Finally, optical methods for the detection of some of the analytes reported in the literature using molecular-imprinting technique have been summarized. These studies show the role of molecular-imprinting technology in food safety, environmental monitoring, and biomedical applications. In short, this chapter describes a brief discussion on molecular-imprinted polymers and their applications in the detection of various analytes.

References

Agrawal, H., A. M. Shrivastav, and B. D. Gupta. 2016. Surface plasmon resonance based optical fiber sensor for atrazine detection using molecular imprinting technique. *Sens. Actuators B* 227: 204–211.

Amabilino, D. B. and J. F. Stoddart. 1995. Interlocked and intertwined structures and superstructures. *Chem. Rev.* 95: 2725–2828.

Athikomrattanakul, U., N. G. Eichelmann, and F. W. Scheller. 2011. Thermometric sensing of nitrofurantoin by noncovalently imprinted polymers containing two complementary functional monomers. *Anal. Chem.* 83: 7704–7711.

Chen, L., S. Xu, and J. Li. 2011. Recent advances in molecular imprinting technology: Current status, challenges and highlighted applications. *Chem. Soc. Rev.* 40: 2922–2942.

Chen, L., X. Wang, W. Lu, X. Wu, and J. Li. 2016. Molecular imprinting: Perspectives and applications. *Chem. Soc. Rev.* 45: 2137–2211.

Dong, J., N. Gao, Y. Peng et al. 2012. Surface plasmon resonance sensor for profenofos detection using molecularly imprinted thin film as recognition element. *Food. Control.* 25: 543–549.

Dvorakova, G., R. Haschick, K. Chiad, M. Klapper, K. Mullen, and A. Biffis. 2010. Molecularly imprinted nanospheres by nonaqueous emulsion polymerization. *Macromol. Rapid. Commun.* 31: 2035–2040.

Gupta B. D., A. M. Shrivastav, and S. P. Usha. 2016. Surface plasmon resonance-based fiber optic sensors utilizing molecular imprinting. *Sensors* 16: 1381.

Kugimiya, A., J. Matsui, T. Takeuchi et al. 1995. Recognition of sialic acid using molecularly imprinted polymer. *Polym. Plast. Technol. Eng.* 28: 2317–2323.

Lanza, F., A. J. Hall, B. Sellergren et al. 2001. Development of a semiautomated procedure for the synthesis and evaluation of molecularly imprinted polymers applied to the search for functional monomers for phenytoin and nifedipine. *Anal. Chim. Acta* 435: 91–106.

Lehn, J. M. 1995. *Supramolecular Chemistry*. New Jersey: Wiley-VCH Publications.

Murase, N., S. I. Taniguchi, E. Takano, Y. Kitayama, and T. Takeuchi. 2016. A molecularly imprinted nanocavity-based fluorescence polarization assay platform for cortisol sensing. *J. Mat. Chem. B* 4: 1770–1777.

O'Shannessy, D. J., B. Ekberg, and K. Mosbach. 1989. Molecular imprinting of amino acid derivatives at low temperature (0°C) using photolytic homolysis of azobisnitrile. *Anal. Biochem.* 177: 144–149.

Ramstroem, O., L. I. Andersson, and K. Mosbach. 1993. Recognition sites incorporating both pyridinyl and carboxy functionalities prepared by molecular imprinting. *J. Org. Chem.* 58: 7562–7564.

Sari, E., R. Üzek, M. Duman, and A. Denizli. 2016. Fabrication of surface plasmon resonance nanosensor for the selective determination of erythromycin via molecular imprinted nanoparticles. *Talanta* 150: 607–614.

Schirhagl, R. 2013. Bioapplications for molecularly imprinted polymers. *Anal. Chem.* 86: 250–261.

Sellergren, B., C. Dauwe, and T. Schneider. 1997. Pressure-induced binding sites in molecularly imprinted network polymers. *Macromolecules* 30: 2454–2459.

Shamsipur, M., A. B. Seidani, J. Fasihi, and H. Sharghi. 2010. Synthesis and characterization of novel ion-imprinted polymeric nanoparticles for very fast and highly selective recognition of copper (II) ions. *Talanta* 83: 674–681.

Shrivastav, A. M. and B. D. Gupta. 2016. SPR and molecular imprinting based fiber optic melamine sensor with high sensitivity and low limit of detection. *IEEE J. Selec. Top. Quantm. Electron.* 22: 6900207.

Shrivastav, A. M., S. K. Mishra, and B. D. Gupta. 2015a. Fiber optic SPR sensor for the detection of melamine using molecular imprinting. *Sens. Actuators B* 212: 404–410.

Shrivastav, A. M., S. K. Mishra, and B. D. Gupta. 2015b. Localized and propagating surface plasmon resonance based fiber optic sensor for the detection of tetracycline using molecular imprinting. *Mat. Res. Exp.* 2: 035007.

Shrivastav, A. M., S. K. Mishra, and B. D. Gupta. 2015c. Surface plasmon resonance based fiber optic sensor for the detection of ascorbic acid utilizing molecular imprinted polyaniline film. *Plasmonics* 10: 1853–1861.

Shrivastav, A. M., S. P. Usha, and B. D. Gupta. 2016a. Fiber optic profenofos sensor based on surface plasmon resonance technique and molecular imprinting. *Biosens. Bioelectron.* 79: 150–157.

Shrivastav, A. M., S. P. Usha, and B. D. Gupta. 2016b. A localized and propagating SPR, and molecular imprinting based fiber-optic ascorbic acid sensor using an in-situ polymerized polyaniline–Ag nanocomposite. *Nanotechnology* 27: 345501.

Shrivastav, A. M., S. P. Usha, and B. D. Gupta. 2017. Highly sensitive and selective erythromycin nanosensor employing fiber optic SPR/ERY imprinted nanostructure: Application in milk and honey. *Biosens. Bioelectron.* 90: 516–524.

Subrahmanyam, S., S. A. Piletsky, and A. P. F. Turner. 2002. Application of natural receptors in sensors and assays. *Anal. Chem.* 74: 3942–3951.

Usha, S. P., A. M. Shrivastav, and B. D. Gupta. 2017. A contemporary approach for design and characterization of fiber-optic-cortisol sensor tailoring LMR and ZnO/PPY molecularly imprinted film. *Biosens. Bioelectron.* 87: 178–186.

Verma, R. and B. D. Gupta. 2013a. Optical fiber sensor for the detection of tetracycline using surface plasmon resonance and molecular imprinting. *Analyst* 138: 7254–7263.

Verma, R. and B. D. Gupta. 2013b. Fiber optic SPR sensor for the detection of 3-pyridinecarboxamide (vitamin B3) using molecularly imprinted hydrogel. *Sens. Act. B* 177: 279–285.

Wackerlig, J. and P. A. Lieberzeit. 2015. Molecularly imprinted polymer nanoparticles in chemical sensing—Synthesis, characterization and application. *Sens. Actuators B* 207: 144–157.

Wang, B., Y. Wang, H. Yang, J. Wang, and A. Deng. 2011. Preparation and characterization of molecularly imprinted microspheres for selective extraction of trace melamine from milk samples. *Microchim. Acta* 174: 191–199.

Whitcombe, M. J., M. E. Rodriguez, P. Villar, and E. N. Vulfson. 1995. A new method for the introduction of recognition site functionality into polymers prepared by molecular imprinting: Synthesis and characterization of polymeric receptors for cholesterol. *J. Am. Chem. Soc.* 117: 7105–7111.

Wulff, G. 1995. Molecular imprinting in cross-linked materials with the aid of molecular templates—A way towards artificial antibodies. *Angew. Chem. Int. Ed. Engl.* 34: 1812–1832.

Wulff, G., R. Kemmerer, J. Vietmeier, and H. G. Poll. 1982. The preparation of chiral cavities in synthetic polymers. *Nouv. J. Chim.* 6: 681–687.

Wulff, G., W. Vesper, E. R. Grobe, and A. Sarhan. 1977. Enzyme-analogue built polymers: On the synthesis of polymers containing chiral cavities and their use for the resolution of racemates. *Macromol. Chem.* 178: 2799–2816.

You, A., Y. Cao, and G. Cao. 2016. Colorimetric sensing of melamine using colloidal magnetically assembled molecularly imprinted photonic crystals. *RSC Adv.* 6: 83633–83667.

Zhang, W., X. W. He, Y. Chen, W. Y. Li, and Y. K. Zhang. 2011. Composite of CdTe quantum dots and molecularly imprinted polymer as a sensing material for cytochrome c. *Biosens. Bioelectron.* 26: 2553–2558.

Zhang, Y., D. Song, L. M. Lanni, and K. D. Shimizu. 2010. Importance of functional monomer dimerization in the molecular imprinting process. *Macromolecules* 43: 6284–6294.

7

Summary and Future Outlook

In summary, the book provides the comprehensive prospects of the optical sensing field including history, fundamentals, research, and applications for the environmental monitoring and biomedical diagnostics. We have mainly focused on the fundamentals of the recognition and transduction units of sensors. The book contains seven chapters including the present one. Chapter 1 summarizes the introduction of sensors and the need of sensors in the field of biomedical diagnostics and environmental applications. Sensors are divided into three main categories: biosensors, gas sensors, and chemical sensors and have been discussed in detail. The chapter also includes the fundamentals of sensors, components of a sensor, and characterization of the sensors based on transducing mechanism in accordance with the recognition schemes. Various parameters which decide the sensor performance have been elaborated in the chapter. Finally, the first chapter ends with the overview of the book. The chapter helps the researcher to understand the basics of the sensors and their characterization such as physical, chemical, and biosensors. Chapter 2 focuses on the understanding of various optical transducing mechanisms for sensing applications. There are several transducing mechanisms such as surface plasmon resonance (SPR), localized surface plasmon resonance (LSPR), lossy mode resonance (LMR), fluorescence, and interference, which have been discussed in the chapter. Among these, SPR, LSPR, and LMR have been elaborated in detail. Further, in the chapter, several highly sensitive new transducing mechanisms such as fano SPR, magneto-optic SPR, and gap SPR have been briefly discussed which need more development. This enables a broad field for the researchers to develop highly selective and sensitive sensors for the industrial, environmental, and biomedical applications. Chapter 3 describes the implementation of optical fiber substrate for the sensing applications. The optical fiber substrate has several advantages such as compact design, light weight, immunity to the electromagnetic interference along with the online monitoring and capability of remote sensing. Further, the optical fiber provides the fabrication of low-cost sensing devices for real-time measurement in harsh environments. Various fiber parameters which affect the performance of the fiber-optic sensors have been discussed. The possibilities of different types of fiber-optic substrates such as straight uniform core, tapered core, and U- and D-shaped probes have been elaborated in detail in the chapter. Chapter 4 is based on the incorporation of nanotechnology in the sensing applications. The chapter deals with the fundamentals of nanotechnology and various physical, chemical, and biological methods for the fabrication of metallic nanostructures. The role of single nanoparticle as the recognition unit and as the transduction unit has also been elaborated in the chapter. The chapter also covers a transduction mechanism, surface-enhanced Raman spectroscopy (SERS) for sensing applications. Further, the role of nanostructure as the recognition unit through various interactions with the sensing analytes has been mentioned. Finally, the chapter discusses, briefly, some of the nanotechnology-enabled sensors and various possibilities on developing the more sensitive sensors using nanostructures instead of metallic thin films. This is because of more surface-area-to-volume ratio, easy accessibility of the analyte to the binding sites, more stability, and the capability of playing the dual role of transduction and recognition unit. Chapter 5 of the book presents the semiconductor metal oxides (SMOs)-based optical sensors for the detection of several environmental parameters and biomedical/clinical analytes. SMO has numerous advantages such as stability, compatibility with various organic/inorganic materials, optical transparency and conductivity in visible and infrared regions, and mechanical flexibility. The chapter covers the role of SMO as the recognition unit and/or as transducer unit for the detection of various gases. To understand the sensing mechanism, various interactions of SMOs with the gases have been elaborated. The interaction of oxidizing and reducing gases by means of adsorption or chemical interaction with SMOs changes the surface chemistry of metal oxides having surface and native oxygen defects. This changes the effective refractive index of the metal oxide and has been detected by means of many transduction methods such as SPR, LSPR, and LMR. The feasibilities and advantages of incorporating the nanostructured SMOs in sensing over the bulk-layered SMO have also been outlined. Since SMO has been the major component of gas sensor probe, a detailed discussion on both the SPR- and LMR-based SMO mechanism has been included. In the last section, a brief discussion on the use of nanostructured SMOs in realizing the biosensors has been provided. Although it has been observed that SMOs are broadly used for the fabrication and designing of the sensors, there is always a possibility to improve the performance of SMO-based sensors by tuning their optical and electrical properties. SMOs and their nanostructures can play several roles such as transducing unit, recognition unit, and immobilization matrix for the fabrication of sensors. Hence, to improve the performance of sensors, the area of SMO-based sensors is widely open for the researchers. Chapter 6 of the book is based on the development of new type of sensing unit called molecular imprinting. The technique includes the freezing of template molecules in a polymeric medium and creating their complementary imprints at their positions in that polymeric medium. The snapshot of the templates in the polymer matrix works as the highly selective recognition sites for the template molecules. Thus, molecular-imprinted polymers (MIPs) are showing their potential in sensing applications by replacing the naturally occurring receptors such as enzymes, antibodies, nucleic acids, and DNAs. In addition, natural receptors also suffer from the disadvantages such as high cost, low-chemical stability, low-binding kinetic, storage problems, and complex immobilization. The chapter covers the introduction of MIP, synthesis steps, and its role in sensing applications. In addition, some of the optical sensors utilizing molecular imprinting for food safety, environmental monitoring, and biomedical applications have been discussed. The MIP technique has a bright future in the development of the cheap and highly selective sensors. There are several challenges in the preparation of MIPs such as imprinting of large-size molecules, preparation of uniform-sized MIP nanostructures, and selection of suitable amount/type of various elements. Thus, there is good scope for the researchers to fabricate the highly sensitive and selective optical sensors for the environmental monitoring and biomedical applications.

Index

G

H

I

J

K

L

M